Introduction to SCIENTIFIC AND TECHNICAL COMPUTING

Introduction to SCIENTIFIC AND TECHNICAL COMPUTING

Edited by
FRANK T. WILLMORE • ERIC JANKOWSKI
CORAY COLINA

CRC Press is an imprint of the
Taylor & Francis Group, an informa business

CRC Press
Taylor & Francis Group
6000 Broken Sound Parkway NW, Suite 300
Boca Raton, FL 33487-2742

Printed on acid-free paper
Version Date: 20160525

International Standard Book Number-13: 978-1-4987-4504-8 (Paperback)

Library of Congress Cataloging-in-Publication Data

Names: Willmore, Frank T., author. | Jankowski, Eric, author. | Colina, Coray, author.
Title: Introduction to scientific and technical computing / Frank T. Willmore, Eric Jankowski, and Coray Colina.
Description: Boca Raton : Taylor & Francis, CRC press, 2017. | Includes bibliographical references and index.
Identifiers: LCCN 2016006492 | ISBN 9781498745048 (alk. paper)
Subjects: LCSH: Engineering--Data processing. | Science--Data processing. | Research--Data processing.
Classification: LCC TA345 .W545 2017 | DDC 502.85--dc23
LC record available at http://lccn.loc.gov/2016006492

Visit the Taylor & Francis Web site at
http://www.taylorandfrancis.com

and the CRC Press Web site at
http://www.crcpress.com

Contents

Foreword

In the relatively brief history of electronic computers—roughly seven decades—we have seen exponential increases in capability and reductions in cost and size, enabled by tremendous advances in technology. These exponential rates of change in capability and cost-efficiency produced the PC revolution, and now the "post-PC" revolution of "smartphones" and other smart mobile devices, and are now driving us toward an "Internet of Things" in which computing devices of all types will be ever-present in our homes, cities, transportation, and perhaps even clothes and bodies. It is easy to understand how computing could now be taken for granted. Computing devices are now pervasive; they are not only on our desks and in our laps, but in our cars and appliances, in our pockets, and even on our wrists. Most people interact with computers daily, hourly, or even constantly for enjoyment or for productivity, without ever really pausing to wonder how they work. In fact, the term "computer" isn't even applied generally to most of the computers in our lives. This is perfectly fine, for the most part. However, computers are also important instruments of science and engineering, necessary for helping us make new discoveries and innovations. For scientists, engineers, and other technical professionals in universities, labs, and companies, computers are increasingly powerful tools for research and development (R&D). Sometimes, very powerful computers—often called "supercomputers"—make headlines for their role in enabling groundbreaking research. Mostly, however, scientific computing systems are out of the public awareness, even more invisible than the pervasive devices we use daily.

The tremendous success of computing in supporting our businesses, running our infrastructure, and enriching our lives has created a huge demand for computing professionals in companies that provide these products and services. The business and consumer markets need experts in cloud computing, data analytics, web services, mobile applications, devices, networking, and so on. The explosion in business- and consumer-focused computing during the past two decades—since the World Wide Web made computing access desirable to everyone—has dominated the market for programmers, and thus changed the focus on computer science programs. The days of scientific computing skills and technologies being the foundation of computer science curricula are long gone, even though computing is also more important than ever in science and engineering. The majority of computer science curricula has understandably evolved with markets for computing technologies. In 2016, essentially all college and graduate students can take courses for programing in Java or for understanding web services, but few can take science-focused programming classes to learn the new version of Fortran or how to use C/C++ for scientific applications. Fewer still can take classes to learn how to use OpenMP and MPI to create high-performance scientific and engineering applications that run on parallel computing systems.

For these reasons, I created and started teaching a course on parallel computing for scientists and engineers when I worked at the San Diego Supercomputer Center (SDSC) in the late 1990s, and taught it at both the University of California, San Diego (UCSD) and SDSU. The class was based on a 2 to 3 day workshop we offered to SDSC users, most of whom were faculty and postdocs who needed supercomputers in their research but had no formal education on how to use them. Shortly after I founded the Texas Advanced Computing Center (TACC) at The University of Texas (UT) in 2001, we started teaching a similar class at UT—and then discovered the need for additional, more basic scientific computing classes on programing in Fortran and C, performance optimization, visualization, data management, and other scientific computing skills. At TACC, we were at the forefront of computational science, and we volunteered our time to share our expertise by teaching classes so students could learn these skills, which were not addressed by the computer science curriculum. Our "scientific computing curriculum" courses were electives for students, but they increased in popularity as students in science and engineering programs at UT realized the need for such skills and understanding to pursue their degrees, especially in graduate research. The classes filled a gap in the curriculum for future scientists and engineers in training, and our course slides filled the gap normally addressed by textbooks.

I am pleased to see that the curriculum pendulum is starting to swing back in the United States, with more universities now offering scientific computing classes: sometimes in science or engineering departments, sometimes in computer science departments, sometimes in new scientific computing/computational science programs. Public appreciation of the importance of computational science also seems to be on the upswing. The advent of the "Big Data" era has impacted science as well as business; data-driven scientific discoveries such as the Higgs boson and confirmation of gravitational waves have made mainstream media headlines, as have research advances in medical research. The White House "National Strategic Computing Initiative," announced in June 2015, has helped to increase public awareness of the need for "exascale" computing systems for new scientific discoveries, industry competitiveness, and national security. However, the general shortage of scientific computing classes at universities remains, as does the scarcity of high-quality, up-to-date textbooks. I am thus very proud that one of the very talented TACC (former) staff who volunteered his time to teach these skills has led a team to prepare a comprehensive, up-to-date scientific computing textbook. For this effort, Frank Willmore deserves far more appreciation from scientists and engineers than he is likely to receive, as is usual for a textbook author. Nonetheless, I have high hopes that this textbook will enrich the education of many students, and enable many new classes to be taught at universities nation- and worldwide. Better trained scientists and engineers will use computing more effectively to do better science and engineering, producing discoveries and innovations in research, industry, and national security for which we will all benefit—even if we don't all appreciate it, all the time.

Jay Boisseau
Vizias Computing
Austin, Texas

Preface

In the fall of 2014, my coeditors Coray Colina, Eric Jankowski, and I organized a group of sessions on the topic of software engineering for the molecular sciences at the annual meeting of the American Institute of Chemical Engineers. Soon thereafter, we received an inquiry from the publisher asking if we would be interested in producing a book on the topic. I had previously taught a course on Scientific and Technical Computing at the University of Texas at Austin, and it had been on my mind to produce a book suitable to accompany that course. My approach was to give my students at least a taste of proficiency in a variety of topics and to find 20% of the information on key topics that were used 80% of the time. In editing this volume, we set out to recruit authors who were both sufficiently knowledgeable in an area and proficient communicators; people who were passionate about these topics and who could explain them well. We asked these authors to write their responses from the following perspective: "If you could give a junior researcher only a single lecture or chapter of notes on a topic, what you would tell them?"

There are two ways in which you can read this book: (1) Cover to cover, and you will gain enough knowledge to be a well-rounded computational researcher and will have a good idea of what the computing terrain looks like and will know where to turn; or (2) Choose chapters as a quick-start guide for various technologies as you discover you need them. When the time comes to go further, follow the references as a starting point. If you are already an enthusiastic scientist or engineer and don't have a lot of experience coding, this should at least be a quick guide to the tools that your colleagues use and give you enough of a taste that you can speak with them intelligently about computation.

This is a technology book, and the process of composing it involved a poignant use of technology. Some chapters were written using MS Word, some with LaTeX, and others with Google Docs. The journey of writing the book has involved the challenges of managing a project team across three continents and several time zones. Documents were shared using Google Drive and meetings were held using Skype. Technology, being a universal constant in a world that does not share a common language, traditions, and so on, brings us together.

In any new endeavor, we struggle to find guides, ostensible mentors, as well as known wizards—people and resources who can and will help us to accomplish what we set out to do. I owe the deepest debt of gratitude to my team of colleagues at the Texas Advanced Computing Center. In particular, Luke Wilson and John Lockman were my copilots the two times I taught the course, and I relied heavily on their depth of experience. A shout out to Yaakoub El Khamra for his never-ending generosity both at work and in contributing his chapter on OpenMP; Carlos Rosales for helping me find answers to the many technical questions; Robert McLay for taking me ever deeper into systems configuration; and John Cazes for keeping me on track through it all. Finally, I thank my coeditors and contributors for believing in my project and bringing this book into existence. I could not have done it without you!

Frank T. Willmore
Level Zero Design

Editors

Frank T. Willmore, PhD, became interested in chemistry and computers growing up in Oak Park, Illinois. He quickly augmented his chemistry knowledge and apparatus beyond his toy chemistry set by regular trips to the public library to find additional books and shopping at local hardware and drug stores for chemicals. He became interested in computers after purchasing a book on video game programming at a fifth-grade book fair. He enjoyed writing his own games and would rewrite the same space invaders–like game several times before purchasing a cassette recorder to save his work. He learned 6809 assembly code so that he could make his games run faster.

Dr. Willmore graduated with a BS in chemistry with minors in math and physics from Valparaiso University (Indiana) in 1993, and went on to pursue an MS in chemical engineering at Iowa State University, where his simulation-based research in catalysis was published in *Physical Review E*. He learned to program in C on IRIX and Ultrix systems and went on to work in software for the next 4 years. He worked for several start-ups, including Evite.com, where he implemented new features and mined user data.

In 2003, Dr. Willmore returned to graduate school to study molecular modeling and simulation and published four more papers, completing a PhD in 2006 under the direction of Isaac Sanchez at The University of Texas. Upon graduating, he paused his research career to found an LLC and built three energy-efficient houses and, in 2009, completed a five-star (highest rating) green home in Austin, Texas. With the completion of that project, he returned to research, completing postdoctoral work at the National Institute of Standards and Technology and joining the Texas Advanced Computing Center as research staff.

Eric Jankowski, PhD, joined the Materials Science and Engineering Department at Boise State University (Idaho) in January 2015. The overall goal of his work is to leverage thermodynamics for societal good. This means understanding the factors that govern molecular self-assembly and using that knowledge to engineer materials for generating energy, storing data, or curing disease. The approach taken by his lab is to create and use computational tools that efficiently generate important configurations of molecules. The goals of these computational models are to (1) provide fundamental insight into material structure when physical characterization is inadequate, and (2) to identify the most promising material candidates when there are too many choices.

Dr. Jankowski earned a PhD in chemical engineering from the University of Michigan in 2012, where he developed computational tools to study the self-assembly of nanoparticles. These tools leverage graphics processors to accelerate computations and provide insight into systems of both theoretical and practical importance. He began focusing on renewable energy generation during his postdoctoral positions at the University of Colorado and the National Renewable Energy Laboratory. At these postdocs, he applied techniques developed during his thesis to understand factors that determine the ordering of molecules in organic solar cells. Dr. Jankowski also enjoys cycling and an ancient board game (go) and can easily be convinced to discuss how themes of efficiency and combinatorics overlap between these hobbies and his professional interests.

Coray Colina, PhD, is a professor in the Department of Chemistry at the University of Florida with an affiliate appointment in Materials Science and Nuclear Engineering. She obtained a PhD degree in chemical engineering at North Carolina State University, working with Keith E. Gubbins, and was a postdoctoral research associate in the Department of Chemistry at the University of North Carolina at Chapel Hill, working with Lee Pedersen. Dr. Colina was previously a faculty member at Simón Bolívar University, Venezuela, and joined the Department of Materials Science and Engineering at The Pennsylvania State University as an associate professor in 2007. Her group strives to understand and predict structure—property relations in functional materials, such as polymeric membranes, hydrogels, biomolecules, and alternative ionic liquids. They use a variety of simulation techniques to gain further understanding of these systems by providing unique insight into structural aspects and phenomena. Complementary to experimental investigations, their work is helping to analyze and interpret experimental results, as well as to predict performance of new materials to guide future experimental design efforts.

Contributors

Brian C. Barnes
U.S. Army Research Laboratory
Weapons and Materials Research Directorate
Aberdeen Proving Ground, Maryland

Victor Eijkhout
Texas Advanced Computing Center
The University of Texas
Austin, Texas

Yaakoub El Khamra
Texas Advanced Computing Center
The University of Texas
Austin, Texas

Chris Ertel
K&E Dagital
Houston, Texas

Todd Evans
Texas Advanced Computing Center
The University of Texas
Austin, Texas

Christopher R. Iacovella
Department of Chemical and Biomolecular Engineering
Multiscale Modeling and Simulation (MuMS)
Vanderbilt University
Nashville, Tennessee

Ahmed E. Ismail
Department of Chemical Engineering
Statler College of Engineering and Mineral Resources
West Virginia University
Morgantown, West Virginia

Haoqiang Jin
NASA Advanced Supercomputing Division
NASA Ames Research Center
Moffett Field, California

Christoph Klein
Department of Chemical and Biomolecular Engineering
Multiscale Modeling and Simulation (MuMS)
Vanderbilt University
Nashville, Tennessee

Paul Kwiatkowski
GroupRaise.com, Inc.
Houston, Texas

Charles Lena
Department of Chemical Engineering
The University of Texas
Austin, Texas

Ryan L. Marson
Department of Chemical Engineering
Biointerfaces Institute
The University of Michigan
Ann Arbor, Michigan

Janos Sallai
Institute for Software Integrated Systems (ISIS)
Vanderbilt University
Nashville, Tennessee

Erik E. Santiso
Department of Chemical and Biomolecular Engineering
North Carolina State University
Raleigh, North Carolina

Paul Saxe
Materials Design, Inc.
Angel Fire, New Mexico

Michael S. Sellers
Bryan Research and Engineering, Inc.
Bryan, Texas

İnanç Şenocak
Department of Mechanical and Biomedical Engineering
Boise State University
Boise, Idaho

Jerome Vienne
Texas Advanced Computing Center
The University of Texas
Austin, Texas

1

Operating Systems Overview

Erik E. Santiso

CONTENTS

Most of the computing resources available to run large simulations use Unix-like operating systems. This is not a coincidence, as the Unix™ paradigm has proven to be ideal for managing systems with a large number of users and includes powerful native tools to manage and process complex dataset. In this chapter, we will introduce some of the basic features of Unix-like systems (which include Linux and the Mac OS X Port, Darwin). This chapter is not meant to be a comprehensive guide or a system administrator guide, but an introduction covering many of the tools that you are likely to use on a regular basis. Much more information on additional tools may be found at the end of the chapter [1,2].

1.1 Introduction: The Kernel, Shells, Processes, Daemons, and Environment Variables

All Unix-like operating systems have at their core a program called the *kernel*. The kernel is the first program that is loaded into memory upon booting and is in charge of all the low-level communications between running processes and the hardware, as well as controlling basic functions such as memory management, process management, scheduling, file management, input/output, and network management. The kernel has full control over everything that happens in the system, and it must run continuously while the computer is in operation.

An instance of running a program is called a *process*, and it is identified in Unix-like systems by a unique number called a process identifier (PID). Note that a program itself is not a process, but rather a file containing a collection of instructions to be executed. The same program may be run multiple times (by the same or different users), giving rise to multiple processes. A special process is the init process, which is executed by the kernel and is in charge of starting and managing other processes. The init process has a PID of 1, and subsequent processes are numbered sequentially as they are created. A *file* is a collection of data contained in any storage media (such as a hard drive or optical disk). Files can also be abstractions for system devices (such as a drive or a network card) and communications between processes.

There are a couple of special types of process that have special names: daemons and shells. *Daemons* are processes running in the background that provide different functionalities to the operating system and do not usually interact with users. For example, the http daemon enables a web server to respond to web requests (such as a remote computer requesting to load a web page). The system-logging daemon enables the writing of system messages to files, which can be used to diagnose potential problems. Usually, daemon processes have names ending with the letter "d"; for example, the http daemon is usually called httpd, and the system-logging daemon is syslogd. There are, however, exceptions to this rule; for example, the daemon in charge of scheduling processes is called cron, and the daemon that provides e-mail services is usually called sendmail.

Shells are processes that enable the interaction between a user and the operating system. A shell provides a set of commands that allow a user to do things like running other processes and managing files. The particular set of commands is not unique, as there are several different shells available in most Unix-like systems. Some of the common shells are as follows:

- csh/tcsh: The C shell (csh) and its more modern successor, tcsh, are designed to use a syntax akin to the C programming language. Although some operating systems use tcsh as their default shell, it is not the easiest to use for beginners and it is a suboptimal choice for scripting (discussed in a later chapter). The C shell has been largely replaced by Bash (see below) in most systems.
- sh/Bash: The Bourne shell (sh) and its modern successor, Bash (Bourne-again shell), are designed to allow scripting, which will be covered in a later chapter. The Bourne shell was the original Unix™ shell. Most modern Unix-like systems use Bash or another modern successor of sh called the Almquist shell (ash).

Other shells include the Korn shell (ksh), which is based on the Bourne shell but includes features from the C shell, and the increasingly popular z shell (zsh), a modern script-oriented shell that incorporates features from Bash, ksh, and tcsh. The rest of this chapter will cover Bash commands, as this is the most widely used shell as of this writing.

Users often interact with Unix-like systems (especially remote systems) via text-based interfaces, such as xterm (in a Unix-like system) or the terminal application (in Mac OS X). Many applications have been designed to run via such interfaces and are usually called *console applications*. Console apps lack a sophisticated graphical user interface (GUI), which may make them unattractive at first, but as a result, they have fewer hardware requirements, are simpler to use, and often allow the user to carry out complex tasks quickly. Some common console apps are pine/alpine (a text-based e-mail client), vim, and emacs (widely used text editors). Most popular scientific software packages, including molecular simulation software, are designed to run from a text-based interface.

Every running process in a Unix-like system (including the shell) has access to a set of *environmental variables*. These variables are often used to store user preferences or to communicate data between processes. A list of all the defined environmental variables can be obtained by running the env command. Some common environmental variables are as follows:

- PATH: This variable contains a list of default paths that the shell searches when a command is invoked. For example, the env command mentioned above actually runs a program that is (usually) located in the /usr/bin directory. Without a PATH variable, in order to run this command, we would have to type the whole absolute path to the executable, /usr/bin/env. Instead, if the PATH variable contains /usr/bin as one of its paths (it usually does), the shell will automatically run /usr/bin/env when env is invoked.
- HOME: Contains the location of the user's home directory. This directory contains the files that belong to the current user, including configuration files. By default, when a shell is invoked, it automatically navigates to the user's home directory unless a user configuration variable specifies otherwise.
- PS1: This sets the appearance of the *prompt* in Bash. The prompt is the text that precedes every command that the user types. For example, if PS1 is set to "[\u \W]$," the prompt will display the user name followed by the current directory, between the square brackets and followed by the "$" sign and a space.
- TEMP: Contains the default location for temporary files.

To see the current value of a particular environmental variable, you can use the echo command, discussed in a later section. For example, $ echo PATH will output the contents of the PATH environmental variable (in Unix-like systems, a "$" in front of a variable name causes the variable name to be replaced with the value stored within; running "$ echo PATH" would simply output the word PATH):

```
$ echo $PATH
/usr/bin:/bin:/usr/sbin:/sbin:/usr/local/bin
```

In Bash, environmental variables can be defined using the export command (which makes the variable accessible to subprocesses), or just by typing the variable name followed by the equal sign "=" and the value of the variable. For example, for user "erik," setting PS1 to "\u rules:" would turn the prompt to "erik rules:":

```
$ whoami
erik
$ PS1="\u rules: "
erik rules:
```

Some environmental variables are defined in user configuration files, which are stored in the user's home directory and are run every time the shell is invoked. These files often have names prefixed with a dot ".", which makes them invisible unless the user explicitly asks for a list of hidden files with ls -a (see the ls command below):

```
$ ls -a
.
..
.DS_Store
.Trash
.Xauthority
```

```
.bash_history
.bashrc
.cache
.config
.dropbox
.fontconfig
.fonts
.gitconfig
.gnuplot_history
.local
.moldenrc
.octave_hist
.profile
.rnd
.sage
(...)
```

An example is the ".profile" file, which usually sets the values of some environmental variables such as PATH. Other configuration files contain instructions to be executed every time a command is invoked; these files often have names that end in "rc". For example, ".vimrc" is used to define default settings for the vim text editor, and ".vmdrc" can be used to define settings for the visual molecular dynamics (VMD) software. The file ".bashrc" contains commands that run every time the Bash shell is invoked and is often used to define user-specific environmental variables and shortcuts to various commands (see the `alias` command in Section 1.5).

1.2 Basic Navigation and File Manipulation

All Unix-like systems share a similar *filesystem*, which provides a set of conventions for storing and retrieving files. The filesystem provides a tree of directories, which themselves contain files or other directories. The directory at the lowest level is the *root directory* and is denoted by a slash "/". The full location of a file of directory starting from the root directory is called the *path* (or the *absolute path*) to that file or directory. For example, the absolute path to the `env` command mentioned in the previous section could be "/usr/bin/env". The absolute path for a command in a particular system can be found using the `which` command, discussed later:

```
$ which env
/usr/bin/env
$ which ls
/bin/ls
```

Paths can be also defined relative to the current directory; in this case they are called *relative paths*. Note that absolute paths are unambiguously defined, but relative paths depend on the current directory. This can be a source of error when running scripts. For example, if the user is currently in the directory /home/erik, the relative path "docs/research" would correspond to the absolute path "/home/erik/docs/research". The current path can be retrieved using the pwd command, mentioned later:

```
$ pwd
/home/erik
$ ls docs/research
funding
in_progress
ideas
papers
(...)
$ ls/home/erik/docs/research
funding
in_progress
ideas
papers
(...)
```

The directory structure can be navigated by using the cd command. Running cd followed by a path will change the current directory to that path. The full path can be omitted if the target directory is within the current directory. There are two common shortcuts used with cd: running "$ cd .." will navigate to the directory containing the current directory, and running "$ cd ~" (or just "cd") will navigate to the current user's home directory (defined by the HOME environmental variable). When the shell is invoked, the current directory is set to the user's home directory.

A typical Unix-like filesystem contains a few default directories under "/", which fulfill different functions at the operating-system level (see Figure 1.1). Some common ones include the following:

- /bin, /sbin, /usr/bin, /usr/sbin: Contain basic shell commands such as cd and env. Usually, /bin (or/sbin) contain commands that should be available to the root superuser (the system administrator, who has permission to run any commands and access/modify any file in the system).
- /lib, /usr/lib: Contain libraries that are used by the commands in the directories above.

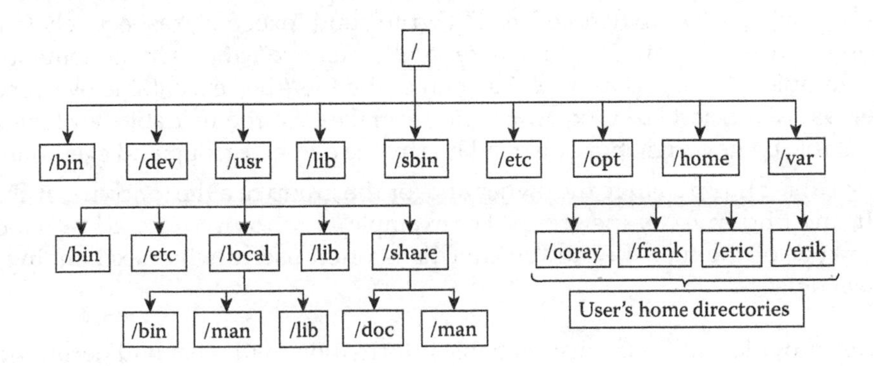

FIGURE 1.1
A diagram showing some of the first few levels of a Unix-like filesystem.

- /etc (sometimes also /usr/etc): Contains system configuration files, including options for initializing daemons.
- /dev: Files that are abstractions for system devices are stored in this directory. This includes hardware abstractions (for example, a file may refer to a hard drive or a network card) and software abstractions (for example, /dev/null, the *null device*, which discards all data sent to it).
- /opt: Contains optional applications.
- /home: Contains user home directories (in Darwin, it is called /Users).
- /var: Usually contains temporary data, such as log files or data used by current running processes.
- /usr, /usr/local: Contain files (including applications) that are available to all users.

These are only some of the possibilities, and different systems may include all or only some of them. Most often users do not need to access or modify most of these directories and mostly interact with files under their own home directory.

In Unix-like systems, all files and directories have individual sets of *permissions* and *attributes*. Permissions define who can read, modify, and/or execute a file/directory. In the case of directories, "read" means listing the contents of the directory, "write" means creating/renaming/deleting files or directories within it, and "execute" means entering the directory (with `cd`) and running commands on files in the directory. These permissions are defined at the user level, the group level, and the universal level (groups are subsets of users that are usually defined in the /etc/group file). For example, a file may have universal read permission (meaning every user can read the contents of the file) but only user-level write and execute permissions (meaning only the owner of the file can modify or execute the file). The permissions and the owner of the file can be modified by the following commands:

- `chmod` is used to modify the access permissions ("mode") of a file or directory. The command can be used with two different syntaxes: one encodes the set of permissions using an octal number (for more information on this version, see the manual page for `chmod`, invoked with "`$ man chmod`"). The other syntax uses the symbols "u", "g", and "a" to denote "user", "group", and "all", respectively; the letters "r", "w", and "x" to denote "read", "write", and "execute", respectively (among others); and "+", "–", and "=" to denote "add", "remove", and "set" (among others). For example, "`$ chmod a+x file`" makes the file "file" executable by everyone, whereas "`$ chmod u=rwx,go=r`" `file` makes the file readable, writable, and executable by its owner, but only readable by the owner's group and everyone else.
- `chown` and `chgrp` change the owner and/or the group of a file (`chown` can change both, and `chgrp`, only the group). For example "`$ chown erik file`" changes the owner of the file "file" to "erik". This is often useful when exchanging files between users.

Attributes provide additional customizable information to files beyond permissions. For example, a file can be *immutable, undeletable, append only,* and many others depending on the particular version of the filesystem. Users rarely need to change these attributes, but if needed (and available) they can be modified with the `chattr` command (`xattr` in Mac OS X).

The most frequently used commands in a Unix-like system are typically those related with managing files and navigating the filesystem. We have already mentioned the `cd` command, which changes the current directory. The other most common ones include the following (the options for each command are not all-inclusive; for more details, refer to the manual page for the command):

- `mv` is used to move or rename files. To rename, use "`$ mv old_name new_name`" (new_name can include an absolute path, in which case the file is also moved to that path). To only move the file, use "`$ mv file path`". For example, "`$ mv file..`" moves the file called "file" to the directory just above the current directory. The path can be relative or absolute.
- `cp` is used to copy files. The syntax is very similar to mv: using "`$ cp old_name new_name`" creates a duplicate of the file named "old_name" and calls it "new_name". Using "`$ cp file path`" will create a copy of the file named "file" at the given path.
- `ls` lists the contents of a directory. If used with no arguments, it lists the contents of the current directory, whereas `ls path` will list the contents of the directory at the given path. This command has a lot of useful options; for example, "`$ ls -a`", mentioned previously, will list all files (including hidden files starting with "."), and "`$ ls -l`" will list not only files but also their permissions, owner, group, size, and date and time they were last modified. The permissions in this case are listed using a string of characters that can be "-", "d", "r", "w", and "x". The first character is "d" for a directory, and "-" for a file, whereas "r", "w", and "x" are used to denote read, write, and execute permissions, respectively. These are listed in the order user/group/everyone. For example, "-rwxr-xr—" denotes a file that can be read, modified, and executed by its owner, read and executed (but not modified) by its group, and only read by everyone else, and "drwxrwxrwx" denotes a directory fully accessible and modifiable by everyone.
- `rm` can be used to delete files. Use this command carefully! Unix-like systems do not typically have a mechanism for undeleting files: if they have not been backed up elsewhere, they are permanently gone. The command has several different options: `rm -d` will delete directories, whereas `rm -r` will recursively delete the whole directory tree below the given directory. Sometimes, depending on file attributes the `rm` command will ask for confirmation to delete the file, this can be avoided by running `rm -f` (be very careful in particular with this option).
- `mkdir` creates an empty directory, whereas `rmdir` deletes an empty. The `mkdir` command will give an error if the directory already exists; this can be turned off by using `mkdir -p`. The `rmdir` command will fail if the directory is not empty, in order to delete a nonempty directory use `rm`.
- `pwd` returns the current directory, as mentioned above.
- `cat` outputs the entire contents of a file. Invoked as "`$ cat file`", it will output the contents of the file named "file" to the terminal. The output can be (and often is) redirected (see input/output redirection below). Note that a binary file (as opposed to a text file) will produce unreadable output.

- `more` and `less` are similar to `cat` in that they output the contents of a file but they stop after every page until a key is pressed. The `less` command has a few additional features, allowing for scrolling text line-by-line, going back a page, and clearing the screen after finishing. In many modern systems, `more` is a synonym for `less`.
- `touch` changes the "last modified" and "last accessed" date and time of a file. If the date and time are not given, it changes both times to the current time.
- `du` outputs the amount of disk space used by a file or directory. When used on a directory, by default, it will recursively list the space used by every file within that directory before giving the total. This can be a lot of output! The maximum depth within the directory can be controlled with the "-d" option. By default, the size is listed in "blocks," which can be defined by the environmental variable BLOCKSIZE (by default, the block size is usually half a kilobyte). Alternatively, `du -k`, `du -m`, and `du -g` will list file sizes in kilobytes, megabytes, and gigabytes, respectively (the "–g" option may not be available in old systems).
- `df` displays disk usage information. This is useful to keep track of how much disk space is available for files. In some systems, the `quota` command can be used to find the amount of disk space available to a particular user and how much of it is being used.

Files and directories can be created by running various commands or programs. One useful way to create and store data on files, as well as using data on a file as input to a command, is *input/output redirection*. Some basic redirection operands in Bash are the following:

- ">": redirects the standard output* of a command to a file. For example, the command "`$ ls -l > filelist`" will write the contents of the current directory (including permissions, etc.) to a file called "filelist" in the current directory.
- ">>" has the same effect as ">", with the difference that, if the target file exists, ">" will overwrite it, whereas ">>" will append the output to it. If the file does not exist, ">>" will behave exactly as ">".
- "2>": redirects the standard error (instead of standard output) of the command to the given file.
- "|" is the "pipe" command, which sends the standard output of a command to another command. This allows the user to chain commands without needing to store intermediate information in files. Piping will be discussed in more detail in Section 1.5.

One particular use of redirection is to create an empty file. This can be done in Bash using ":>"; that is, `:> filename` will create an empty file called "filename."

Finally, a useful shortcut when manipulating files is *tab autocompletion*. In most Unix-like systems, the "tab" key can be used to complete the name of a file or directory when typing a command. This saves a lot of time, especially when file names are long. For example, consider a directory containing three files named "file11", "filewithaverylongname", and

* Commands in Unix-like systems produce two separate output streams: standard output and standard error. The former is the usual output of the program, whereas the latter contains error messages. This will be discussed in more detail in later chapters.

"file111". Instead of typing "cat filewithaverylongname" to list the contents of that file, it is enough to type "cat filew" and then the "tab" key. Note that typing "cat file" and then "tab" will not achieve this result because there is more than one file with a name starting with "file". Autocompletion in such a case will complete up to the point where there is more than one option. For example, typing "cat file1" and then "tab" will autocomplete "file1" to "file11" and then wait for user input, since there are two possibilities starting with "file11".

You may be wondering why would one want to use long file names. The reason is similar to the one for using long variable names when programming: long names provide more information about the contents of the file and can be used to systematically identify files in a hierarchy. It is always a good idea to be as explicit and systematic as possible when naming files and directories, not only for the purpose of readability but also to make scripting easier (this will be discussed in more detail later).

1.3 Interacting with Remote Computers

More often than not, when running simulation packages, you will be running them on a remote computer (usually a computing cluster/supercomputer). This subchapter will cover some of the different ways to interact with remote computers.

There are several different protocols that can be used to interact with a remote computer. At the most basic level are protocols that allow the user to run shell commands on the remote computer and protocols that allow exchanging files between different computers. The most basic example of the first is the `telnet` command, which allows a user to start a shell session in a remote computer. The simplest example of the second protocol is `ftp` (file transfer protocol), which allows for downloading/uploading files from/to a remote server. Both `telnet` and `ftp` were widely used in the early years of Unix-like systems but have been largely superseded by `ssh` and `scp/sftp` for security reasons: both commands transmit data, and particularly passwords, across the network in plain text, that is, without any encryption. This means that an intruder monitoring network traffic could have access to user passwords and data. For this reason, in many modern Unix-like systems, these commands are disabled.

As a result of the security issues with `telnet` and `ftp`, more modern connection protocols were developed to achieve the same results. The most widely used of these are based on *RSA encryption*, named after Ron Rivest, Adi Shamir, and Leonard Adleman, who developed the method. This encryption method allows users and computers to freely exchange information across a network while ensuring that only the intended recipient can read the information. The RSA cryptosystem uses a pair of passwords, or "keys," that are separately used for encryption and decryption. The encryption key for a particular user/computer is visible to everyone (for this reason, it is often called the "public key"), whereas the corresponding decryption key is known only to the user/computer (and is called the "private key"). RSA works as follows: imagine that a user called Alice wants to send a file to the user called Bob. Alice will encrypt the file using Bob's public key, which is visible to everyone, and send the encrypted file to Bob. To decrypt the file, Bob uses his private key, which is known only to him. RSA can also be used to "sign" files, enabling Bob to verify that Alice was indeed the originator of the message. The mathematics behind RSA is fascinating but is beyond the scope of this chapter. The short story is this: RSA is based on the difficulty of finding prime factors of very large numbers, for which there is no known efficient algorithm.* The public key is derived from a large number that is the product of two large primes, and the two primes define the private key.

The RSA-based equivalent of telnet is the *secure shell* or `ssh`. This command allows a user to start a shell session on a remote computer. For example, if user "erik" has an account on the remote computer "unity.ncsu.edu", the user can connect to the remote computer by running "`$ ssh -l erik unity.ncsu.edu`" (or, alternatively, `ssh erik@unity.ncsu.edu`). This will work if the local computer uses a Unix-like system (including Mac OS X). For Windows, a third-party program that allows for ssh connections is needed; examples include PuTTY, Bitvise, MobaXterm, and Tera Term. To end a connection, use the `exit` or `logout` command.

In addition to `ssh`, the `scp` (secure copy) command can be used to directly exchange files between different computers. The syntax of `scp` is the same as that of the cp command, but the file names of the local or remote file must include the user and host name of the computer where the file is located. For example, to transfer the file named "localfile" in the current directory to the account of the user "erik" on the computer "unity.ncsu.edu", the corresponding `scp` command could be as follows:

```
$ scp localfile erik@unity.ncsu.edu:./mydirectory
```

This would copy the file to the directory "mydirectory" under the home directory of the user erik (the "." after the colon is short for the home directory; alternatively, the full path could be given). The same syntax can be used to retrieve a file from a remote computer; for example, the following command would copy the file with the absolute path/home/erik/myfile on the remote computer unity.ncsu.edu onto the current directory on the local computer:

```
$ scp erik@unity.ncsu.edu:/home/erik/myfile.
```

The "." at the end is short for the current directory. Alternatively, a path (relative or absolute) to copy the file to can be given.

An alternative to `scp` is the `ssh` file transfer protocol, or `sftp`, which is a RSA-based replacement to the old ftp protocol. Unlike `scp`, the `sftp` protocol allows for running additional commands on the remote computer, including some of the basic commands discussed in the previous section. When connected to a remote computer through `sftp`, the "get" command can be used to retrieve a remote file to the local computer.

In Windows systems, where the `scp` and `sftp` commands are not available, a third-party client can be used to achieve the same functionality. Examples include WinSCP and Filezilla. PuTTY, as well as the other clients mentioned above, also include `scp` and `sftp` tools (called `pscp` and `psftp` in the case of PuTTY).

The `ssh`, `scp`, and `sftp` commands will require the user to enter its password every time to access the remote computer. It is possible to set up `ssh` keys in order to allow for passwordless login and file transfer. The `ssh` keys (based on RSA) are used to identify "trusted" computers on the remote server. These keys are stored in the .ssh directory within the user's home directory (note that, as the directory name starts with a dot, it is

* There are theoretical algorithms, such as Shor's algorithm, that can factor large numbers into primes efficiently, but they require quantum computers and are not currently feasible.

invisible unless listed with ls –a). The procedure to set up passwordless login in a remote server is as follows:

1. On the local machine, create the key pair that will identify it to the remote machine (this needs to be done only once). This can be done using the `ssh-keygen` command. This will (by default) create a file called id_rsa under the ".ssh" directory.
2. The contents of the file .ssh/id_rsa file created in the previous step must be appended to a file on the remote server called authorized_keys, under the remote ".ssh" directory. This can be done, for example, by copying the .ssh/id_rsa file to the remote server using scp and then using cat and redirection with ">>" to append the data. For example, if the remote computer is unity.ncsu.edu and the remote username is "erik," the commands would be

   ```
   $ scp ~/.ssh/id_rsa erik@unity.ncsu.edu:./
   ```

   on the local computer, and then

   ```
   $ cat ~/id_rsa >> .ssh/authorized_keys
   ```

   on the remote computer. After that, the file ~/id_rsa on the remote computer can be deleted.

After this is done, using `ssh`, `scp`, or `sftp` on erik@unity.ncsu.edu from the local computer would not require entering a password.

In addition to the protocols described above, there are more specialized ways to interact with remote computers. For example, files that are available via a web server can be downloaded using the `wget` command. For example, the file "index.html" at the URL http://www.ncsu.edu can be downloaded to the current directory by running

```
$ wget http://www.ncsu.edu/index.html
```

This is useful when retrieving files such as installers from web sites.

1.4 Regular Expressions

When setting up simulations and/or analyzing and processing data, we often wish to run commands on multiple files, possibly many files. Sometimes, we wish to search for and extract information from one or multiple files. Instead of writing the same command over and over again, *regular expressions* allow us to search for, modify, and execute files (or content in files) whose names follow a pattern. Regular expressions are one of the most powerful tools in Unix-like systems and can save a lot of work.

If you have worked on a terminal, you have probably encountered a few regular expressions. A common use is to use the asterisk character "*" to match any string of characters. For example, if you want to delete all files in a directory with a name starting with "file," you can use the command "`$ rm file*`" to do it. The "*" in a regular expression means "any group of zero or more characters," so that command would match a file called "file," as well as a file called "file_00" but *not* a file called "1file". The "*" is only one of several special symbols (called *metacharacters*) that can be used in regular expressions. In this section, we will cover some of the most common ones (by the way, the usage of "*" in the shell is a special case).

Most of the examples in this chapter will involve searching for text within a file using regular expressions. A common way to do this is to use the `grep` command. This command searches for a particular string, or for strings that match a regular expression, within a file and outputs the lines in the file that match. Let's consider a file called "example," containing the following lines:

Name	Age	Phone	Address
Alice	31	(919)123-4567	1021 Third St., Raleigh, NC
Bob	33	(617)515-5540	124 Willow St., Cambridge, MA
Jenny	21	(310)867-5309	315 W 83rd St., Los Angeles, CA
Mary	22	(971)405-1532	3110 Hood St., Sandy, OR
Matthew	41	(919)315-3282	1150 SW Cary Parkway, Cary, NC

Running the command "`$ grep NC example`" would list the following lines:

Alice	31	(919)123-4567	1021 Third St., Raleigh, NC
Matthew	41	(919)315-3282	1150 SW Cary Parkway, Cary, NC

We can use this file to test several possible regular expressions. First, let's assume we want to find lines that start with the letter "M". We can do this by typing "`$ grep '^M' example`". This would return the following:

Mary	22	(971)405-1532	3110 Hood St., Sandy, OR
Matthew	41	(919)315-3282	1150 SW Cary Parkway, Cary, NC

The character "^" is an example of an *anchor character.* Anchor characters denote particular positions within a line. The most commonly used anchor characters in regular expressions include the following:

Anchor Character	Match
^	Beginning of a line
$	End of a line
\<	Beginning of a word
\>	End of a word

For example, in the file above, "`$ grep 'A$' example`" would match lines ending on the letter A (the entries for Bob and Jenny), and "`$ grep 'y\>' example`" would match lines containing words that end with the letter y (the entries for Jenny, Mary, and Matthew).

Since some characters in regular expressions have special meanings, we need a convention for cases when we actually want to search for those characters. For example, if you tried to search for the dollar sign using "`$ grep $ example`", the command would return all lines in the file (because all lines have an end). In the shell, this is resolved by preceding the special character with a backslash "\", which acts as an escape character. Therefore, to search for the dollar sign, the correct syntax is "`$ grep '\$' example`" (which, in our case, would return nothing since there are no dollar signs in the file).

In addition to anchors, there are other useful special characters that can be used in regular expressions. One of these is the dot, ".". In a regular expression, a dot matches any single character. For example, using the file above, "`$ grep '\<2.\>' example`" would match lines that contain words starting with the number 2 and followed by a single character. This would return the entries for Jenny and Mary. As above, to actually search for a dot, we need to escape it with a backslash.

Additional flexibility can be obtained by using character lists and ranges. These can be defined by using square brackets. The inside of the square brackets may contain a list of all the characters to match and/or a range of characters (in numeric or alphanumeric order). For example, the regular expression [aeiou] will match any vowel, whereas the expression [a-z] will match any lowercase letter. Ranges and lists can be combined; for example, the regular expression [aeioul-9&] would match any vowel, number between 1 and 9, or an ampersand (&). Using a caret (^) inside the square brackets would match any character *except* the ones listed. For example, the regular expression [^a-d0-9] would match any character that is *not* a number or one of the lowercase letters a, b, c, and d. Of course, anchors and ranges can be combined, for example, with the file above, "`$ grep '[A-Ca-z]$' example`" would return lines for which the last character is an uppercase A, B, or C or a lowercase letter (i.e., the title line and the entries for Alice, Bob, Jenny, and Matthew).

In addition to character lists, regular expressions allow the use of modifiers indicating the number of times an expression should be repeated. Two common such modifiers are the asterisk "*" and the question mark. When an asterisk is placed after a character or character list/range, it will match *zero or more* times that expression. A question mark will cause it to match *zero or one* times that expression. Note that zero is included in the range, so if the pattern is absent, it will also be a match. Note that, in the shell, an asterisk actually behaves as the regular expression ".*"; that is, it will match zero or more of any character.

In addition to "*" and "?", it is possible to use modifiers to specify a particular number of repetitions, as well as a range. This is done using "\{" and "\}". If there is only one number between the two, it specifies a number of repetitions. For example, running the command "`$ grep '[aeiou]\{2\}' example`" would match any line containing two consecutive vowels. In our case, this would return the entries for Alice (who lives in Raleigh) and Mary (who lives on Hood St.). If there are two numbers separated by commas between "\{" and "\}", the first number denotes the minimum and the second the maximum number of repetitions. If the comma is included but one of the numbers is omitted, that indicates that there is no minimum or maximum. For example, running "`$ grep '[a-zA-Z]\{8,\}' example`" would return lines that contain at least eight consecutive letters, that is, only the entry for Bob (since Cambridge has nine letters).

The metacharacters discussed above are used in basic regular expressions. *Extended regular expressions* include additional functionality. Two such common extensions are the "+" repetition modifier and the *or* operator "|". The plus sign, "+", acts in a similar way to "*", but it matches the previous expression *one or more* times instead of zero or more. The *or* logical operator is useful to match one or more from a particular list of entries. For example, the expression 'Alice|Bob' would match the entries for Alice and Bob in our example file. Note that `grep` does not normally support extended regular expressions by default, but they can be enabled by using the `-E` flag (or by using `egrep`, which is equivalent to `grep -E`).

There is more to regular expressions than discussed here, and there are particular sets of regular expressions that work with some programs and not others. More information on regular expressions can be found in the references listed at the end of the chapter.

1.5 Other Commonly Used Commands

In this section, we will list some commands that are commonly used in Unix-like systems, or utilities that are sometimes useful for specific tasks. This is intended to be a reference list only, and many options for each command are not discussed. Much more information is available in the manual pages for the respective commands, which can be invoked using `man` (see below).

- `alias`: Allows the user to define a name for commonly used long commands so that they do not need to be typed every time. `alias` is often used in user configuration files (such as .profile or .bashrc) to define custom commands.
- `aspell`: An interactive spell checker. Can be quicker than pasting the text into a Word processor for checking.
- `bc`: This is an arbitrary-precision calculator. It allows for doing quick calculations in the shell and can be useful when very high precision is needed when calculating things within a script. However, for complex tasks, a utility such as octave or sage may be more appropriate (see Section 1.9). As an example, the command to calculate the exponential of the square root of 3.5 in bc from the shell would be

  ```
  $ echo 'e(sqrt(3.5))' | bc -l
  6.49367543608378672613
  ```

  This is an example of piping commands, which will be discussed in Section 1.6.
- `cal`: Returns a calendar for the current month. Also includes options to return calendars for other months, to find holidays, and for displaying Julian calendars.
- `clear`: Clears the terminal screen.
- `cut`: Allows for "cutting" pieces of each line in a file. This is useful to extract data from files where the data are organized in columns (very often the case in output files from simulation packages). Most commonly, `cut` is used with the –d and –f options, which indicate the character that delimits the different fields and the field(s) to print. For example, to print columns 3 and 5 of a file containing entries separated by commas, the corresponding command would be "`$ cut -d, -f 3,5 file`". To print all columns from 2 to 5 (including them), use –f 2-5. An alternative way to extract data from files that many people prefer over `cut` is to use the `awk` language (see Section 1.8).
- `date`: This command displays (or sets) the current date and time. Setting the date is allowed only by the system administrator.
- `diff`: The `diff` command shows differences between two files. It is very useful for version tracking, since it can tell whether two versions of a file are different and what are the differences. `diff` will list lines that have changed between the first and second file and indicate a change with the letter "c" (for example, 2c2

means the second line in the first file is a changed second line in the second file). If a line has been deleted, it uses "d" and lists the original line, whereas a line that has been added is listed with "a."

- `echo`: Repeats ("echoes") anything following the command. This is the simplest output mechanism used in scripts and is also sometimes useful to pass strings to other commands using pipes/redirects. By default, `echo` inserts a newline character at the end of the output (i.e., it scrolls to the next line); this can be turned off with the –n flag.
- `find`: This command searches for files hierarchically within one or more directories (i.e., it keeps navigating all the subdirectories it encounters) and can run several different operations on them. By default, it lists the files it finds, but it can run other commands on them. Regular expressions can be used in conjunction with `find` to locate files with names matching a pattern.
- `finger`: Displays user information. If invoked with no options, it behaves similarly to who (i.e., lists all users logged in). When followed by a username, it lists basic information about that user. It also lists the user's "plan," which can be modified by creating/editing the file ".plan" in the user's home directory.
- `head`, `tail`: These two commands are used to list the initial (`head`) or final (`tail`) lines of a file. By default, list the first (or last) 10 lines. The –n option can be used to control the number of lines listed. For example, "`$ head -n 20 filename`" will list the first 20 lines of the file named "filename." Using a plus sign in front of the number will cause `tail` to list the file starting from that line number until the end (instead of that number of lines). For example, "`$ tail -n +5 filename`" will list all lines starting from the fifth one.
- `gzip`: This is a popular compression utility among Unix-like systems. By default, it will compress a file. With the –d option, it will decompress a compressed file. This format is often used in conjunction with `tar`.
- `ln`: This command creates a link to a file (similar to "aliases" in Mac OS X and "shortcuts" in Windows). Two types of links can be created: "hard" links (the default) actually link to the underlying data. When hard links to a file exist, all links must be deleted in order for the file to be deleted (deleting the original file does not affect the hard link). Instead, "symbolic" links behave in a similar way to shortcuts in other operating systems: they point to the original file but they become invalid if the original file is deleted or renamed. Symbolic links are created using the –s option, and are the most commonly used.
- `logname/whoami/id`: These commands return the identity of the user running it. `logname` and `whoami` return the current user's username. `id` returns additional information such as the numeric user id, group id, and other system-dependent information.
- `mail`, `mailx`: These commands can be used to send e-mail from the shell. It is not a very efficient way to send mail, but it is useful, for example, to have a script notify a user by e-mail automatically when it is done running.
- `man`: Most Unix-like operating systems come with a detailed online documentation system that contains manual pages for commands, as well as other useful operating system conventions and concepts. Invoking `man command` will bring up the manual page for that command. This is one of the most useful commands, especially for beginners. Use it!

- `paste`: This is the opposite of `cut`. It concatenates lines from files. For example, if two files (file1 and file2) contain a column of data each, "`$ paste file1 file2`" will return two columns, the first with the entries from file1 and the second with the entries from file2.
- `printf`: Allows for formatting output. This is useful when writing scripts but can sometimes be useful to create sets of files/directories following a specific format. The syntax is borrowed from the C/C++ command with the same name. `printf` can round numbers, pad them with extra zeros, write them in scientific notation, convert them to hexadecimal and octal, justify text, and many other options.
- `seq`: Generates a sequence of numbers given the first and last number in the sequence and the spacing between numbers (if no initial number is given, it starts from 1). This is very useful for scripting or running multiple commands at once (see Section 1.8 and Chapter 9 for examples). The output can be formatted using `printf` syntax and can be automatically padded with zeros by using the –w option. This is particularly useful when creating a range of files/directories or using any application where it would be desirable for alphabetical and numerical ordering to be equivalent. For example, for files called file1, file2, … file10, running the `ls` command would list file10 after file1, since the default is using alphabetical order. If the numbers are padded with zeros (i.e., file01, file02, … file10), the order will match the numerical order.
- `sort`: As its name indicates, this command sorts the lines of a file. By default, it sorts lines alphabetically. Typical options include –n, which sorts numerically instead, –b, which ignores spaces at the beginning of the line, and –u, which removes repeated entries.
- `tar`: This is a utility for creating "tape archives." It allows for storage of one or many files (including their directory hierarchy) into a single file. This is useful for transferring or backing up sets of files. `tar` is often used in conjunction with a compression utility (usually gzip) to reduce the size of the resulting file. The most common options used with `tar` are –c (create a new tape archive), –r (append additional files to an existing tape archive), –t (list the contents of the tape archive), and –x (extract the files from the archive). Also, the –z option automatically compresses/decompresses the file using gzip. The –v (verbose) option causes tar to list files as it reads/writes them. For example, the command to compress the contents of a directory called `dir` into a gzipped file called `tarball.tgz` would be "`$ tar -czf tarball.tgz dir`" (the –f option tells tar to write to the specified file; otherwise, it writes to standard output). The corresponding command to uncompress would be "`$ tar -xzf dir`".
- `tee`: This command enables simultaneous redirection to a file and output to screen. Normally, when running a command, the standard output of the command goes to the screen. If redirected with ">", the output is written to a file instead of the screen. `tee` allows for doing both, that is, writing standard output to one or more files and to the screen. Typical usage is `command | tee filename`, where "command" is the command to run and "filename" is the name of the target file where standard output will be copied.
- `uniq`: This command can be used to remove or find repeated lines in a file. Without any option, it removes consecutive duplicate lines from a file. With the –d option, it instead returns a list of the duplicated lines. Running `uniq` on the output of `sort` is equivalent to running `sort -u`.

- `units`: This is a very complete interactive unit conversion utility. It is fairly straightforward to use, when run, it will ask "You want:", expecting a unit or quantity with units. Then it will as "You have:" expecting the new units. For example if entering "15 lb" after "You have:" and "kg" after "You want", the program will return 6.8038856, which is the equivalent mass in kilograms. If only units are given, it returns the conversion factors. The program only handles multiplicative conversions; for example, it cannot convert degrees Celsius to Fahrenheit.
- vim, emacs: These are two different text editors. Users often like one or the other. Both are very powerful, and covering all their features is beyond the scope of this chapter, but the reader is encouraged to learn at least one of them, as being able to do advanced text editing in the shell can save a lot of time.
- wc: This is a utility to count lines, words, or characters in a file. The default is to return the number of lines, number of words, and number of characters in the file (in that order). Each can be requested individually using the –l, –c, or –w option (for example, wc –l will return only the number of lines in the file).
- `whereis`: This command can be used to locate the executable file (or files, if there is more than one version/copy) for a program in the filesystem. For example, in most systems, "`$ whereis echo`" will return /bin/echo. Some versions of `whereis` can provide additional information, such as the location of the source code for the command or the location of its manual page.
- `which`: As mentioned in the beginning of the chapter, Unix-like systems use the PATH environmental variable to define locations in the filesystem to search for commands by default. This allows you to just run "`echo`" instead of having to type "`/bin/echo`" every time. The drawback of this is that, if more than one version of a program is available, it may not be obvious to the user which version is invoked by default. The `which` command answers this question. Running `which` followed by a program's/command's name will return the specific location of the command that is run when that command is invoked.
- who: Displays who is logged into the server.
- `xargs`: Many commands in Unix-like systems expect a list of arguments following the command. Sometimes, it is useful to use the output of one command as the list of arguments for another. This can be done with `xargs`. This is different from piping, which sends the output from one command as the input for another. Instead, `xargs` constructs an argument list and runs the program with that argument list as if all the arguments were typed in the command line directly. For example, running `find . | xargs touch` will run the `touch` command on all the files in the current directory and its subdirectories, recursively (`find` returns the list of all such files, and `xargs touch` runs touch with every element of that list as arguments). Occasionally, `xargs` is used to rearrange data. For example, if a file named "filename" contains one entry per line, "`$ xargs < filename`" will return all the entries on a single line. The –n option can be used to limit the number of elements returned per line (for example, `xargs -n 3` will list three arguments per line).
- `yes`: This command is a relic of older versions of Unix™ that may still be useful in some situations. When run, it just repeats the letter "y" (short for "yes") indefinitely. The main reason for its existence is that, sometimes, certain commands will ask the user to confirm actions one by one. For example, when deleting a lot of

files using `rm` (without the –f option, or if the –f option is disabled), the shell will prompt the user to confirm deleting every single file one by one. If instead the output of `yes` is piped to the `rm` command, for example, running "`$ yes | rm *`", then all the prompts will automatically be answered as "yes." Use this at your own risk! The command can also be used to repeat other words by running `yes` followed by the word to repeat.

1.6 Running and Managing Running Processes

Every action in a Unix-like system involves running a program, since shell commands are themselves programs. Processes (i.e., instances of execution of a program) can be run interactively or in the background. Interactive processes expect direct input from the user (if needed) and produce output directly to the terminal (unless redirected). While an interactive process is running, the user cannot run other commands unless the process is moved to the background. Processes running in the background, however, allow the user to run other commands while they are running. Note that, however, unless a utility such as nohup is used, background processes started within a shell will be automatically stopped when the user exits the shell.

Every running process in a Unix-like system uses two output streams, *standard output* and *standard error*. As their name indicates, the standard output stream is used for the program's normal output, whereas the standard error stream is used for error messages. In addition to these streams, processes can communicate with other processes via *signals* and *pipes*. Signals are notifications that processes send when particular events occurred. Often, these events are related to errors that arise while executing the program, such as invalid mathematical operations (e.g., a division by zero), or an illegal instruction, or by the program exceeding maximum file size or CPU time limits. Processes can have defined signal handlers, which receive signals and respond to them accordingly. In some cases, signals can trigger a *core dump*, which means that a copy of the memory of the program is stored on the hard drive. This can be used for debugging purposes.

Signals can be sent to running processes from the shell using the `kill` command. This command identifies processes by their process identifier (PID), which can be found using the process status (ps) command. Signals can be identified by name, or by number. Some of the signals commonly used in managing processes with `kill` include the following:

- Hangup: This signal is identified as HUP (or SIGHUP) or by the number 1. When a terminal window is closed, all processes that were started by that terminal are sent this signal. Usually, this causes those processes to terminate (an exception is daemons). It is possible to avoid having a process terminated by the hangup signal by using the `nohup` command. The name comes from the days of dialing up a computer via a phone line.
- Interrupt: This signal can be identified as INT (or SIGINT) or by the number 2. This signal is sent to a process that was started interactively from a terminal window when the user sends an interrupt request (typically, this is done by pressing Control-C from the terminal).
- Kill: This signal is identified as KILL (or SIGKILL) or by the number 9. The kill signal immediately terminates a running process, and it cannot be caught, handled,

or ignored by the process. This is often used to stop a process that has "hanged" and does not respond to other signals.

- Abort: This signal is identified as ABRT (or SIGABRT) or by the number 6. This signal terminates a program, just like the previous three, but unlike them, it usually generates a core dump. The abort signal is often generated by the running process itself when a runtime error occurs.
- "Alarm clock": This signal can be identified as ALRM (or SIGALRM) or by the number 14. This signal is sent to a process that has a specified time to run, when the time runs out. It usually causes the program to terminate.
- Terminate: This signal is identified as TERM (or SIGTERM) or by the number 15. This is the "nice" way to end a process, as the signal can be caught and it gives the process a chance to perform memory cleanup, write data to disk, and so on. Ideally, it is better to try sending a TERM signal to a process before using the KILL signal.

The `kill` command can be invoked in several different ways. For example, using the –s option followed by the signal name and the PID will send that signal to the named process. Alternatively, the signal name or number can be given, preceded by a dash. All of the following are equivalent:

```
$ kill -s KILL 410
$ kill -s SIGKILL 410
$ kill -9 410
$ kill -KILL 410
```

All of these will result in the process with PID 410 being immediately terminated without any possibility of doing signal handling. An alternative command that can be used to send signals to processes by using the command invoked to run the process instead of the PID is `killall`. It has the advantage of not requiring the user to find the PID using `ps`, but it will send the signal to all running processes invoked with that command by the current user, so it should be used with care.

There are many other possible signals, and covering all of them is beyond the scope of this chapter. A couple of common ones that you may deal with as you write programs are the "floating point error" signal (SIGFPE), which is sent to a process that attempts an invalid arithmetic operation (not necessarily with floating-point numbers!), and the "trap" signal (SIGTRAP), which is sent to processes when an exception occurs.

Processes can also communicate with each other via *pipes*. Pipes allow for "chaining" processes, that is, to send the standard output of one process as standard input to another process. This is one of the most useful and time-saving features of Unix-like systems, as it eliminates the need to create intermediate files or variables and allows for carrying out fairly complex operations using a single line of code. In the shell, pipes are represented by the vertical bar character "|". The command to the left sends its output as input to the command on the right. For example, the following command will list all files in a directory containing the word "potato" in their name:

```
$ ls -l | grep potato
```

This works as follows: the output of `ls -l` (i.e., the line-by-line listing of files in the current directory with additional information on permissions and owner; see Section 1.2) is sent as input to the grep potato command, which lists only those lines that contain the string "potato". Pipes can be chained, allowing for much more complex operations, and are very useful to process output from simulation packages. For example, the NAMD [3] code for molecular dynamics simulation periodically writes output to its log file containing the string "ENERGY:", followed by information such as the current time step, various contributions to the potential energy, the kinetic energy, temperature, and others, separated by multiple spaces. The time step is on the second column, the total energy in the 12th column, and the temperature in the 13th column. If we want to list the values of these three quantities, discarding the first 100 values (which could correspond, for example, to the equilibration phase of the simulation), the following command will do it:

```
$ grep "ENERGY:" file.log | tail -n +101 | tr -s ' ' | \
  cut -d ' ' -f 1,12,13
```

This works as follows: grep picks all the lines from the file containing the string "ENERGY:" (the example assumes the file is called file.log), tail removes the first 100 lines (it starts with line 101), tr removes consecutive spaces (we will discuss tr in more detail in Section 1.8), and cut lists fields 1, 12, and 13 on the resulting text. The use of tr is needed in this example because cut uses only single spaces as delimiters. We will see an easier way to do this using the awk language in Section 1.8. Of course, if we want to store the resulting information into a file instead of printing to the screen (for example, to use in a plotting program), we can redirect the output by writing > outputfilename at the end of the line. The backslash, "\", is used to break the command into multiple lines.

As the example above shows, pipes are a powerful tool for processing data, as well as carrying out complex operations. It pays off to invest time experimenting with pipes.

Below is a list of other commands that are useful for managing running processes.

- ps: This command stands for "process status," and it can be used to list all programs that a user has started via a terminal (including the shell and processes running in the background). Without any option, it typically displays a list of the processes running by the current user, including their PID, an identifier for the terminal associated with the process (TTY), the CPU time the process has used, and the command used to start that process. It is possible to include other information in the output using the –o and –O options, and it is also possible to list processes being run by other users or processes without controlling terminals. ps has a long list of options; you can find an extensive list in its man page.
- top: This command is an alternative to ps that lists all system processes sorted in a particular way (by default, it sorts by PID in decreasing order). In addition to showing running processes in real time, it shows other useful information such as CPU usage and detailed memory usage. The –o option can be used to control how processes are sorted. A common option is –o cpu, which lists in decreasing order of CPU usage. Other common options are –o command, which sorts by command name; –o time, which sorts by execution time; and –o vsize, which sorts by memory size. By default processes are sorted in decreasing order, but this can be changed by adding a plus sign to the corresponding keyword. For example, top -o +cpu will sort by increasing order of CPU usage.
- nohup: As explained above, when a process is started from a terminal, it is sent the "hangup" signal as the terminal is closed (which usually means that the

process is terminated). nohup allows the user to start a process and prevent it from being killed on hangup. To use it, simply write `nohup` followed by the name of the command to run and any arguments needed. Output that would normally go to the terminal gets written to a file called nohup.out in the current directory.

- ./, `source`: To run an executable program that is in a directory not listed in the PATH environmental variable, it is necessary to give the full path to the program. It is very common for users to navigate to the directory where the executable resides and then use a dot and slash "./" followed by the executable name or, alternatively, give the path from the current directory starting with "./". The "./" tells the shell that the program is in the current directory (a single dot is used in Unix-like systems to denote the current directory, and two dots, "..", to denote the parent directory). An alternative is to use the `source` command followed by the executable name.
- &: A process can be run in the background by adding an ampersand, "&", to the end of the line invoking the job. That way, the user can continue running commands as the program is running, instead of having to wait for the program to be done to continue using the terminal. If the output is not redirected, however, the background process will continue to stream its standard output to the terminal, which can be distracting, and thus, the ampersand is commonly used with output redirection. For example, the following will start the program `myprogram` in the current directory and redirect its standard output to file `myoutput`:

```
$ ./myprogram & > myoutput
```

- CTRL-C: This sends the "interrupt" signal to a process that is running interactively. Usually, this is done to stop execution of the program.
- CTRL-Z, fg, bg: A process that has been started interactively can be temporarily stopped, (but not terminated) by pressing CTRL-Z. When this is done, the job is in "suspended" status and can be resumed by using the `fg` or `bg` commands. `fg` will cause the process to be resumed in the "foreground"; that is, it will take control of the terminal until it finished running. Instead, `bg` will resume the process in the background. One instance when this is useful is programs that require interactive input but take a long time to run: in that case, the user can start the program interactively, enter any input that is required, and then use `CTRL-Z` followed by `bg` to send the process to the background.
- `jobs`: This command lists any processes that are suspended or running in the background, starting with an identifying number (usually [1], [2], etc.). To bring a process in this state to the foreground or background, it is possible to use `fg` or `bg` followed by the number preceded by a percentage sign. For example, `fg%1` will cause the process with the number 1 to run in the foreground, whereas `bg%2` will resume process 2 in the background.
- `exec`: This replaces the current shell with a different command. It differs from `source` in that, when the process executed by `exec` ends, the terminal will be closed. This command is often used to replace the current shell or to manage processes within scripts.

- `sleep`: This command can be used to wait for a specific time (in seconds) before running a command. For example, if we want to run program `myprogram` in the current directory starting 2 hours from now, this can be achieved with:

  ```
  $ sleep 7200 ; ./myprogram
  ```

  The semicolon ";" can be used to separate several consecutive commands on the same line.
- `wait`: Sometimes, especially in scripts, we want to wait until a particular process finishes running to start another process. For example, the full output from the first program may be needed by the second one. The `wait` command can be used to achieve this. The syntax is `wait PID`, where PID is the process id of the background process that the user wishes to finish before continuing.
- `watch`: This command can be used to execute a program repeatedly and show the output every time on the screen. This is sometimes useful to monitor a system, or a running program. By default, it updates the output every 2 seconds, but this can be changed with the –n option. The –d option highlights changes from the previous output. For example, to see the contents of a directory change in real time, one can use "`$ watch -d ls -l`". This command is not available in all systems.

1.7 Shell History and Shortcuts

Shells in Unix-like systems often have mechanisms to recall commands that have been previously executed, which saves a lot of time, especially when executing long commands. The `history` command can be used to bring up a list of the most recent commands executed by the current user. The commands are preceded by numbers and sometimes also by the time of execution. In Bash, the history is typically stored in a (hidden) file called .bash_history in the user's home directory, unless the HISTFILE environmental variable is set (in that case, the variable contains the location of the history file). A useful feature of the `history` command is that it enables storing the commands run in a particular session on a given file, which may be useful to recall how you did something. For example, running the command `history -w filename` will copy the current history to the file with the given name. The current history can be deleted by running `history -c`.

Commands in the user's history can be recalled, and this is usually a major time saver. In some systems, this can be done using the `fc` command, which opens a text editor allowing editing the history, and rerunning commands from it. Alternatively, the history can be navigated in a few other ways. In most systems, the up and down arrow keys can be used to bring up commands in the history. The up arrow brings up the previous command, whereas the down arrow brings up the next one (CTRL-P and CTRL-N can also be used instead of the arrow keys). It is also possible to recall commands by pressing CTRL-R and then typing a string that was present in the original command. For example, a command starting with "grep" can be invoked by typing CTRL-Rgrep. If multiple commands contain the string, it is possible to switch from one to another by pressing CTRL-R after typing the string.

The CTRL-R shortcut is often used to recall long commands, and often, the user may want to change something in the command. Although it is possible to navigate the command line using the right and left arrow keys, this moves the cursor character by character and can be slow if the line is long. However, there are shortcuts that can be used to navigate a long command line: CTRL-A and CTRL-E will move the cursor to the beginning and end of the line, respectively, and ALT-right arrow and ALT-left arrow move the cursor one word to the right and to the left, respectively, instead of character by character. These shortcuts can save a lot of time when dealing with long commands.

An alternative mechanism to reuse commands in the history is the *bash bang*. The "bang" refers to the exclamation mark character, "!", which is the first character in most of the shortcuts. The following are common Bash bang shortcuts:

`!n`: This will rerun the command listed as number "n" in the history list (you will probably need to run `history` to bring up the list with the numbers).

`!-n`: This runs the command n lines before the current line. For example, !-1 will rerun the last command executed.

`!!`: This is a synonym for !-1; it reruns the most recent command. This is commonly used in conjunction with `sudo`, if the current user has the ability to run commands as superuser (if you are not a system administrator, you most likely cannot).

`!string`: This will rerun the most recent command starting with the given string. For example, typing `!man` will rerun the last command starting with `man`.

`!string:p`: Sometimes, you may not be sure of which command will run when using the !string shortcut. Adding :p at the end will only print the command that will run, but not execute it.

`!?string`: This is similar to !string, but the string can be anywhere in the command line, not necessarily at the beginning.

`^string1^string2`: This will rerun the previous command but will replace the first occurrence of string1 with string2. This is often used to correct mistakes when typing long commands.

In addition to reusing commands, a very useful feature of the Bash bang notation is the ability to reuse command arguments. The arguments from the previous command can be invoked using the !:n syntax, where n is the nth argument (zero is the command itself). The notation from regular expressions for the first and last commands also works, and thus, !^ gets substituted to the first argument (i.e., it is equivalent to !:1), and !$ gets replaced by the last argument. Also, !* gets replaced by all the arguments, and a range can be specified as n–m. For example, let's say that the previous command run was as follows:

```
$ ls -l file1 file2 directory1
```

If we afterward try running "`$ more !:2`", this will be replaced by "`more file1`" (note that argument number 1 is –l), whereas running "`$ !:0 !^`" will translate to "`ls -l`". Running "`$ cd !$`" will translate to "`cd directory1`" and running "`$ ls !:2-3`" would yield "`ls file1 file2`". In this particular example, there is not a big gain from using the Bash bang, but when the names of the files or directories are long (or include long paths), it saves a lot of time. In practice, the one most often used is !$.

It is also possible to recover arguments from commands prior to the previous one, but the notation is a bit more cumbersome. To refer to the command n lines above, the bang character must be replaced by !-n. For example, to recover the third argument from the command right before the previous one, the shortcut is !-2:3. The characters $, ^, and * also work; for example, !-3$ is the last argument from three commands earlier.

Recovering arguments through Bash bang shortcuts is very often done to avoid typing long file names and/or paths. In order to help with this task, there are modifiers that can be added to Bash bang shortcuts to further manipulate such arguments. The most common are :h, which keeps only the path but removes the file name; :t, which removes the path but keeps the file name; and :r, which removes suffixes (anything after a dot; for example, file.txt will become file). These modifiers can be combined. As an example, if the previous command is

```
$ ls ~/myDirectory/myFiles/file.txt
```

then "`$ ls !^:h/other`" would convert to "`ls ~/myDirectory/myFiles/other`" (useful to refer to different files within a long path), "`$ ls !^:t`" would become "`ls file.txt`", and running "`$ ls !^:r`" would translate to "`ls ~/myDirectory/myFiles/file`". Finally, running "`$ ls !^:t:r`" would translate to "`ls file`". Some of these more complex shortcuts do not get used as often, but they can save a bit of time if you remember them. While they may take time to learn, they allow the user to sustain a heightened state of awareness of their work, rather on automatically typing and retyping.

1.8 Advanced Text Manipulation

Users often encounter situations when they need to carry out fairly complex or repetitive tasks involving text files. For example, we might want to run a lot of simulations where only a single parameter is changed (say, the temperature). Creating all the directories and input files by hand in this situation would be cumbersome, and it would instead be useful to have a mechanism to automatically generate the required files. This would require replacing text multiple times on a "master" file. Another example is processing output from simulation software. Very often, the information we are after is only in a subset of the lines in the output file and needs to be further processed (for example, properties should be averaged and their standard error should be computed). Unix-like systems have tools that allow for all of these tasks to be carried out directly within the shell. We have seen some of the commands that can be used for this purpose in the previous sections. In this section, we will briefly discuss two other very powerful tools: the stream editor `sed` and the `awk` programming language. We will also discuss the character translator (`tr`), which enables additional functionality, especially in combination with other commands.

The stream editor, `sed`, provides a very powerful language to manipulate and modify text. We will not extensively cover all of its functionality, as this is enough to deserve a book by itself [4]. We will, however, cover some of the most common ways to use it.

`sed` is a pattern-matching language. It allows the use or extended regular expressions to find, print, and/or modify text. One of the most common uses of `sed` is to replace text on files. The syntax for this is `sed 's/string1/string2/' filename`. This will replace the first instance of string1 on every line of file "filename" with string2 and write the result to standard output. As usual, if you want to store the result in a file, you can redirect the output with ">".

As an example, imagine that a file called potatoes.txt contains the following lines.

```
One potato, two potatoes,
Three potatoes, four,
Five potatoes, six potatoes,
Seven potatoes, more!
```

Running "`$ sed 's/potato/tomato/' potatoes.txt`" will produce the following output:

```
One tomato, two potatoes,
Three tomatoes, four,
Five tomatoes, six potatoes,
Seven tomatoes, more!
```

Note that sed only replaces the first instance of "potato" on each line and replaces it even if its only part of a word (as in "potatoes"). If the goal is to replace *all* instances of "potato" with "tomato," a "g" (standing for "global") should be added at the end of the sed command. Running `sed 's/potato/tomato/g' potatoes.txt` would produce the following:

```
One tomato, two tomatoes,
Three tomatoes, four,
Five tomatoes, six tomatoes,
Seven tomatoes, more!
```

This functionality of sed provides a very useful mechanism to generate a lot of files quickly. Imagine, for example, that you intend to run 21 molecular dynamics simulations of a system at 21 different temperatures, ranging from 250 K to 350 K in steps of five. Preparing all the input files for this system by hand would be time consuming and boring. Instead, you can prepare a directory to contain all the simulation files as subdirectories and a master file where every instance of the temperature is written as, say, "XXX". Then you can type the following lines from the Bash shell:

```
for i in `seq 250 5 350`
do
mkdir -p temp_$i
sed 's/XXX/'$i'/g' inputfile > temp_$i/inputfile
done
```

The `for` command, which implements one form of looping in Bash, will be discussed in detail in Chapter 9. In this particular example, `` `seq 250 5 350` `` first gets evaluated into the list of numbers from 250 to 350 in steps of five (the backward apostrophes, `...`, cause the command between them to be run as a new process, take its output, and substitute it in). The `for` command then causes the variable `i` to take every value on that list sequentially between the `do` and `done` commands. Two commands are run for every value of `i`: "`mkdir -p temp_$i`" creates a directory with the name "temp_###", where "###" is the current value of `i` (recall that the dollar sign "$" in front of the variable name replaces the name with the value stored in the variable). The –p option avoids an error message if the directory already exists (alternatively, the output of `mkdir` can be redirected to /dev/null). The second command uses `sed` to replace the "XXX" string in the master file (called "inputfile" in this example) with the current value of `i` and redirects the output to a file called "inputfile" in the newly created directory. After running this command, there will be 21 subdirectories called temp_250, temp_255, all the way up to temp_350, and each will contain an input file where the temperature has been set to the desired value. This is much faster than manually creating 21 directories and editing 21 input files!

The example above illustrates the power of `sed` to automate tasks but also shows that it pays off to name your files and directories systematically, as it makes it easier to write scripts to manipulate them.

It is also possible to search and replace conditionally. For example, if we want to replace the word "potato" with "tomato" only in lines that contain the word "four", we can use the command "`$ sed '/four/s/potato/tomato/' potato.txt`". `sed` will then search for lines containing the word "four" and only replace the first instance of "potato" with "tomato" in those lines (in our example, the change will happen only in the second line). You can also match lines that do *not* match the pattern by adding an exclamation mark "!" before the "s".

Furthermore, it is also possible to match regular expressions with `sed` and to modify the resulting strings. In regular expressions, the ampersand, "&" stands for the most recent match to a regular expression. For example, running "`$ sed 's/[aeiou]/&&/g' potatoes.txt`" would duplicate every lowercase vowel in the file:

```
Onee pootaatoo, twoo pootaatooees,
Threeee pootaatooees, foouur,
Fiivee pootaatooees, siix pootaatooees,
Seeveen pootaatooees, mooree!
```

There are more complex ways to use `sed` to search and replace, but they are less common, and often, people who are not `sed` wizards prefer to use a text editor for such tasks, so we will not cover them here. Note, however, that it is possible to make multiple substitutions consecutively by piping the output of one `sed` command to another with "|", which is often useful.

Another useful command in `sed` is "d", which causes a line to be deleted if it matches an expression or a specified range of lines. For example, "`$ sed '/four/d' potatoes.txt`" will delete the second line in our example, whereas "`$ sed '1,3d' potatoes.txt`" will delete lines 1 through 3 (leaving only the fourth line).

Finally, another common use of `sed` is printing a particular range of lines in a file. This can be done using the –n option (which turns off printing every line) and using "p" to request printing lines that match a particular expression or a particular range of lines. For example, using `sed -n '1,3p'` will print lines 1 through 3 in the file, whereas `sed -n '/[FT]/p'` will print lines that contain the letters "F" or "T". Depending on the version of `sed` installed in the computer, it is also possible to skip lines in a file using the tilde "~". For example, "`$ sed -n '3~5p'`" would print every fifth line starting with the third one (this works in GNU `sed`).

Besides `sed`, another very useful command to manipulate and extract information from text files is `awk`. `awk` is a powerful programming language that allows for the processing files line-by-line. As with `sed`, a thorough coverage of all the functionality of `awk` is beyond the scope of this chapter.

A common use of `awk` is printing particular fields (columns) in a document. Often, this can also be done using `cut` (and possibly `tr`), but the syntax in `awk` is much simpler and easier to remember. For example, running "`$ awk '{print $1, $5}' filename`" will print the first and fifth fields on every line of file "filename." Recall the example from Section 1.6 where we wanted to print columns 1, 12, and 13 of every line in a file containing the string "ENERGY:", discarding the first 100 lines. We achieved this with the following command:

```
$ grep "ENERGY:" file.log | tail -n +101 | tr -s ' ' | \
  cut -d ' ' -f 1,12,13
```

With awk, this is shortened to:

```
$ grep "ENERGY:" file.log | tail -n +101 | \
  awk '{print $1,$12,$13}'
```

which is much simpler.

In general, an awk command has the following structure:

```
awk 'BEGIN{<commands>}{<commands>}END{<commands>}' file
```

Each of the three sections within the curly brackets may contain commands with a syntax very similar to that of the C programming language. Commands are separated by semicolons, ";". It is possible for the program to span multiple lines, although for complex tasks requiring many lines of code, it is usually easier to write a script (see Chapter 9). The commands within the BEGIN section are executed once at the beginning of execution, the commands within END are executed once at the end, and the commands within the middle curly brackets are executed once *on every line* of the input file. Numbers preceded by a dollar sign, "$", become the fields on the line, as in the example above. Also, "$0" stands for the entire line. There are also a few variables that have special meaning in awk. Commonly used ones are the following:

- FS (field separator): This variable contains the character or characters that separate the fields on each line. By default, it is whitespace, but it can be set to any character or sequence of characters, including a regular expression. For example, setting `FS='[,;]'` will make commas and semicolons both work as field separators. As with everything else, to recover the value of FS, there must be a dollar sign in front.
- RS (record separator): This variable contains the character or characters that separate the different "lines." By default, it is a newline character. RS is not as commonly used as FS.
- NF (number of fields): Contains the number of fields in the current line. Usually, this is written as $NF, since it is read from the input file, not set.
- NR (number of records): This contains the number of "lines," or records, in the file. As with NF, typically this is read, not set.
- OFS and ORS: These are like FS and RS, but they control the behavior of the `print` command. OFS is the character (or characters) separating the fields in the output, and ORS is the record (line) separator.

In addition to changing the field separator by setting the value of the FS variable, it is also possible to change it from the command line by using the –F option. For example, `awk -F ','` will cause the field separator to be a comma (the quotation marks are optional; `awk -F` would also work).

There are many simple tasks that can be done easily with awk within the shell, which avoids having to transfer files and opening them with different programs. For example, a common task when running simulations is calculating average values. A simple code that does this in awk is

```
awk 'BEGIN{tot=cnt=0}{tot+=$1;++cnt}END{print tot/cnt}'
```

This will print the average of the numbers on the first column on the file. You could pipe the output of the tail command in the "ENERGY:" example above to this and it would return the average energy (replacing the $1 with $12). If you find yourself using a script like this often, you can create an alias to it on your .bashrc file, for example:

```
alias findavg="awk 'BEGIN{tot=cnt=0}{tot+=$1;++cnt} \
END{print tot/cnt}'"
```

You can automate similar tasks, such as calculating standard deviations and standard errors, in a similar way. In addition to simple arithmetic, awk has a number of special functions available, such as trigonometric functions, exponential, logarithm, and square root.

There is much more to awk than what we have covered in this chapter, as it is a fully featured scripting language. However, as mentioned above, for tasks that are much more complex, it is more efficient to use a language like Python, which is covered in a later chapter.

The final command we will mention in this chapter is a simpler one: the character translator `tr`. This command provides the ability to substitute or delete certain characters from a text stream, including special characters such as the newline (represented as \n), carriage return (\r), tab (\t), and others. It can also be used to delete repeated instances of a character with the –s option, as illustrated in Section 1.6. For simple substitutions or deletions, especially those involving newlines and other special characters, `tr` often has a simpler syntax than the sed or awk commands do, which carry out the same tasks. As an example, the command to replace all newlines in a file with colons, ":", would be, using `tr`:

```
tr '\n' ':' < file
```

whereas achieving the same result in awk would require manipulating the record separators:

```
awk 'BEGIN{RS="\n"; ORS=":"}{print $0}' file
```

which is much longer.

1.9 Useful Packages

One of the best features of Unix-like systems is the availability of free, useful software tools. There are far too many to cover in a single chapter, but a few of them that are often useful to analyze data from simulations are listed below. This is a very limited list, and there are many other packages with similar functionality.

- gnuplot: This is a command-line-driven plotting package. It allows for quick visualization of results from simulations, without the need for copying the output files to a different location to open with a separate plotting application. gnuplot can be used to produce high-quality graphics and has a relatively simple language. The easiest way to get started is to run the `help` command within gnuplot, but there are a number of online resources and books to help [5].
- `xmgrace`: This is another plotting tool that enables the user to interactively modify its output.
- `octave`: Octave is an open-source math package based on MATLAB®.
- `sage`: This is a very complete open-source math package, including tools for numerical and symbolic calculations. It incorporates the functionality of many other packages, including the NumPy, SciPy, and Sympy libraries from Python, as well as that of Maxima, R, and the ATLAS library. It uses a language based on Python, which will be discussed in Chapter 9.
- `vmd`: Standing for "Visual Molecular Dynamics", vmd is a molecular visualization package that includes a lot of features to analyze simulation results. It can be scripted using the `tcl` language.
- `openbabel`: This is an open-source set of packages that enable interconversion between many different file formats typically used to store information from molecular simulations. It has a simple command-line syntax but also provides a library for developing computational chemistry software.

References

1. S. Powers and J. Peek, *Unix Power Tools*, 3rd Ed., O'Reilly, Sebastopol, CA, 2002.
2. UNIX Tutorial for Beginners, http://www.ee.surrey.ac.uk/Teaching/Unix/index.html.
3. J.C. Phillips et al., *J. Comp. Chem.* 26, 1781 (2005).
4. A. Robbins and D. Dougherty, *sed & awk*, 2nd Ed., O'Reilly, Sebastopol, CA, 1997.
5. For example, P.K. Janert, *Gnuplot in Action*, or L. Phillips, *gnuplot Cookbook*.

2

Machine Numbers and the IEEE 754 Floating-Point Standard

Frank T. Willmore

CONTENTS

You are a student of science, you have faith in the good people of the world. You have been told that computers represent everything with zeros and ones, and you have faith that the engineers of such machines have taken appropriate measures to make sure that everything will work out all right when you write programs and give and share information in decimal format. Indeed, you have run spreadsheets and simple codes and are reasonably sure that you are in fact getting the results you expect. But you also know that limitations exist and that no machine can remember a number with infinite precision; things have to break down somewhere. And if you are a good scientist, perhaps this has bothered you a little bit—both not knowing where things break down and not knowing what you do not know. In this chapter, we discuss what you have always wanted to know about machine numbers, whether or not you knew it.

Consider as a simple example the byte `10110110`. I assume you know that eight bits is a byte, and that a bit is zero (`0`) or one (`1`). Because each bit is in either of these two states, and because there are eight bits in the byte, there are $2^8 = 256$ possible sequences of eight bits. If you don't believe me, try writing out different combinations. One very straightforward use of one byte is to store the value of a number. The simplest encoding represents the nonnegative integers from 0 to 255. It would be an easy choice to say that `00000000` represents 0, `00000001` represents 1, `00000010` represents 2, and so on. A slightly fancier encoding of the bits in a byte allows us to represent negative integers: We could say that if the first bit is zero, that means the number is negative, for example, and that −1, the negative integer closest to zero, would be encoded by `10000000`. Depending on the choice of how numbers are encoded into bits, operating on those numbers and interpreting those numbers can vary drastically. In order to avoid the headaches that arise from different encodings, the computing community has created encoding standards.

Before standardization, vendors would supply their own rules for storing numbers. The Institute of Electrical and Electronics Engineers (IEEE) 754 floating-point standard emerged in 1985 to reconcile these rules in the best interest of those involved. This chapter deals specifically with the 2008 revision of the IEEE 754 standard and its implications. At the time of writing, a revised 2011 standard has been accepted, yet most microprocessors currently manufactured are designed to adhere to the 2008 standard, so this will be the focus, and all info here will remain relevant.

2.1 Integer Types

First, a word on integer types. Integer types are the simplest machine representation of the counting numbers and if signed (+/–) can also represent negative integers. The simplest integer type would be some sort of Boolean value, equal to either zero or one. The underlying representation and size of this and other integer types can vary extensively across languages, language standards (C90 versus C99, etc.) and platforms, as well as 64 versus 32 bit modes, which is unfortunate. Since the smallest addressable unit of memory is a byte, it is reasonable to assume that the size of this 1-bit type will be at least a byte.

This is by no means intended to be a complete treatise on the treatment of integer types, merely a warning. You may not always get from your compiler the integer representation you think you asked for, and what you think you are asking for may not even exist on every platform. Research LP64 versus ULP64* if you really want to know more about the sizing of integer (and pointer) types. As one example, the "short" and "long" modifiers to the int type are intended to give longer or shorter integer types as needed, but in reality, they may all be the same size.

2.2 Unsigned and Signed Integers

In memory, unsigned integers are represented as the sum of bits, from most significant byte (MSB) to least significant byte (LSB). The 32-bit integer represented here is 64 + 32 + 16 + 4 + 1 = 117:

2147483648	1073741824	536870912	...	64	32	16	8	4	2	1
0	0	0	...	1	1	1	0	1	0	1

Signed numbers are more complex, with the most significant bit of the most significant byte representing the *sign* of the number. To make things yet more complicated, negative numbers are stored as *two's complement*. To get the two's complement of a number, flip all the bits representing the number and then add 1. It is done this way so that addition is handled the same for both signed and unsigned values, and the machine does all this for

* http://www.unix.org/version2/whatsnew/lp64_wp.html.

you so you need not worry about two's complements unless you are particularly curious. The two's complement of –1 in 32 bits is illustrated below:

Sign	1073741824	536870912	...	64	32	16	8	4	2	1
1	1	1	...	1	1	1	1	1	1	1

Adding 1 to –1 should give 0. Indeed, if you add 1 (`000...0000001` in binary) to `111...1111111`, all the bits turn to 0 (`000...0000000`) because there is no place to store the final `1` that would be carried over. That `1` goes into the bit bucket, which is the programmer's way of saying it is thrown away. Doing the same thing, the unsigned value (`111...1111111`) gives the same result (`000...0000000`) and again the last carry bit goes into the bit bucket. That is, for a 32-bit integer, adding 1 to (2^32 – 1) gives 0. The proper term for this is called *overflow*.

It is a smart move to check the sizes of types you will be using in your code. This can be handled at compile time by more complicated mechanisms, but the following runtime check in C can deliver some peace of mind:

```
assert (sizeof(long long)==8); /* make sure it's 8 bytes */
```

If your code depends on a type being (at least) a certain size, and you plan on possibly representing values that require that many bytes, this will let you know if there is a problem.

2.3 Single versus Double Precision

The IEEE 754-2008 standard describes a representation called binary32. Colloquially, this is still referred to as "single precision." There is also a representation called binary64, colloquially referred to as "double precision." These terms are probably familiar and only slightly misleading. Although it contains twice the information, double precision is arranged slightly differently, and actually provides slightly more that twice the precision, what a bonus!

To begin our discussion of floating-point representations, we will first discuss single precision. We are humans and probably think of numbers like 7, 5.4, 3.33333333, 4/5, "several," and so on, and we can make decisions based on the mathematical meaning we associate with the symbols we see (e.g., whole number, fraction, and repeating decimal). Machines, on the other hand, simply do what they are told. So, if we want the machine to store a number, we first need to tell it which set of rules to use. Let's choose single precision and try to store the number 0.21875. Sounds easy enough. This is what the machine sees:

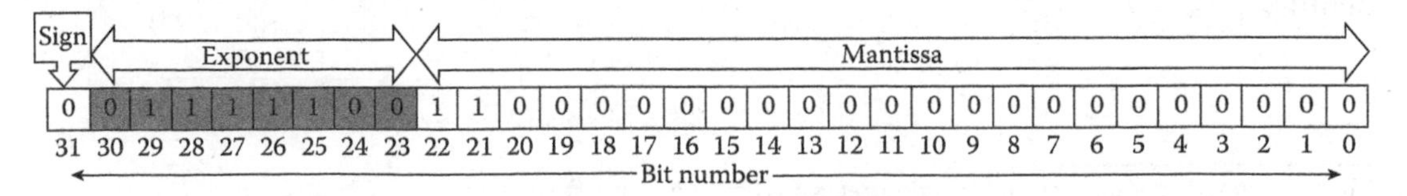

What? Too confusing!? Let's break down and simplify this binary representation to understand what is going on. Notice that there are 32 bits, that is, 32 ones or zeroes. There are 8 bits in a byte (that is just a definition), so this means binary32 or "single precision" uses 32 bits or 4 bytes to store a number. We are all agreed on that! Looking a little closer, we see that there are three groups of bits: the sign, the exponent, and the fraction (aka mantissa, to use a fancy term). The sign seems straightforward, like with integers: We think of numbers as positive and negative, and we use the first bit to say which it is. And it turns out that the standard says that 0 means the number is positive, and that 1 means it is negative. This might not seem intuitive. It is done this way for the same reason as negative numbers are stored two's complement. It is just more convenient for the machine to handle addition and subtraction.

To best understand the exponent and mantissa, think scientific notation, only instead of writing 2.1875×10^{-1}, it all has to be in binary! In scientific notation, we are used to seeing a number written in terms of an exponent and powers of 10, for example, $2.1875 \times 10^{-1} = 2 \times 10^{-1} + 1 \times 10^{-2} + 8 \times 10^{-3} + 7 \times 10^{-4} + 5 \times 10^{-5}$. Interpreting in terms of powers of 2, 0.21875 becomes $1.7500\ldots \times 2^{-3} = (1 \times 2^{0} + 1 \times 2^{-1} + 1 \times 2^{-2} + 0 \times 2^{-3} + 0 \times 2^{-4} + \text{zeros}\ldots) \times 2^{-3}$. A little confusing, but let's not concern ourselves with getting to this representation, only to convince ourselves that it can be done.

So this gives us a set of value of the exponent = −3, and a set of ones and zeroes for the mantissa = `11100`.... To make matters even more confusing, the standard provides that the exponent be stored as an unsigned integer, but with the exponent value of zero being represented by the binary value `01111111` (decimal 127) and positive or negative exponent values counting up or down from there. This is done for machine efficiency, definitely not for clarity to the science student! Under this representation, our exponent value of −3 is represented by the binary `01111100`.

To store the mantissa, the binary number gets shifted left or right (and the exponent decremented or incremented) until its first digit is a one and is in the 2^0's place. The adjacent bits represent 2^{-1}, 2^{-2}, and so on. Since this first bit is always 1, there is no reason to store it. This means that even though our mantissa is `11100`..., the first `1` does not count, so what gets stored is `1100`... Memory was once really expensive, so anytime you could save a bit, you did. There is an awful lot of electrical engineering cleverness tied up in all of this, which is not always obvious to a programmer coming from a different background. The takeaway points to remember are the following:

- There are 8 bits of information allocated to saying where the decimal point should be placed.
- There are (effectively) 24 bits of information allocated to storing the number as precisely as possible.
- There is a single bit allocated to say whether the number is positive or negative.

Additionally, exponent values of `00000000` are used to indicate subnormal numbers (see below) and `11111111` to indicate NaN (not a number, with mantissa nonzero) as well as +/− infinity (with mantissa zero). See the entry for single precision in Wikipedia* for more details.

* https://en.wikipedia.org/wiki/Single-precision_floating-point_format.

2.4 Double Precision

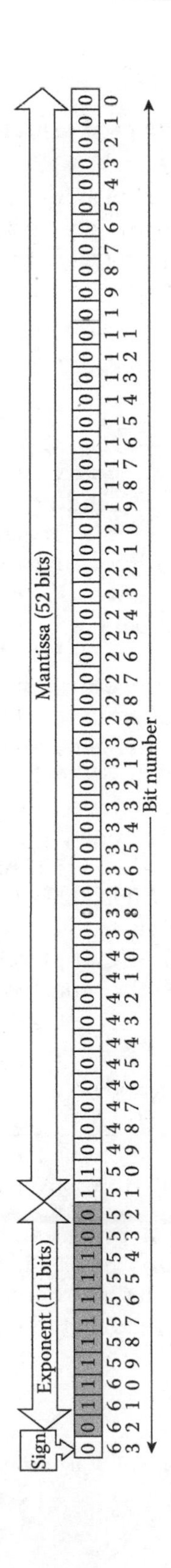

Good news! Double precision* is, in fact, set up very nearly identically to single precision. What is notable are the following:

- There is still only 1 bit for sign.
- The exponent is 11 bits wide, as opposed to 8, allowing for numbers to be 23 times larger or smaller.
- The fraction or mantissa is now 52 bits, instead of 23. There is still one implicit bit, though, giving an actual increase in precision of 53/24, or slightly more than double. So double precision is actually more than twice as precise. I told you!

2.5 Overflow and Underflow

Overflow and underflow for integers are pretty straightforward. You add 1 to the highest possible integer you can get, and you end up with the lowest possible integer and vice versa. For a floating-point number, it is the same idea, but it is now about orders of magnitude, rather than actual magnitude. Increasing or decreasing numbers simply causes the exponent to get higher or lower. Floating-point numbers have a range of exponents available for storage of numbers of various sizes. It is the *exponent* that overflows or underflows!

Many are familiar with the idea of overflow, but what about a number that is too small? Is such a thing even possible? In machine parlance, yes. It is. And the IEEE standard goes to great lengths to try to keep from losing a number, just because it is so small. Above, we stated that the mantissa is shifted and the exponent incremented/decremented so that the most significant bit winds up in the ones' place. When a number is too small, this shifting cannot be done without asking the machine to remember an exponent below its lower limit of –127. What happens in this case is the machine records the exponent binary value of 00000000 (corresponds to 2^{-127}) and stops trying to shift the mantissa. The implicit stored 1 is no longer implicit. This enables the machine to squeeze a few more decimal places out in this case. Why do all this? Basically because there is space left in the binary domain to do something, and this is a convenient feature to include.

2.6 What a Subnormal Number Often Means

Although subnormal numbers have a representation and interpretation, arithmetic operations do not typically return subnormal values, although it is possible to set compiler options to do so when the compiler and hardware support it. Unless this is done explicitly, receiving a subnormal number generally reflects a situation in which memory

* https://en.wikipedia.org/wiki/Double-precision_floating-point_format.

is misinterpreted, for example, an integer being interpreted as a floating point in this C example:

```
main()
{
  // binary representation 00000000000000000000000000000101
  int my_int = 5;
  // interpreting as float gives subnormal
  float my_float = *(&my_int);
  printf("My subnormal value is:%f\n", my_float);
}
```

This gives:

```
My subnormal value is:  7.00649e-45
```

For comparison, the smallest representable *normalized* value is $2^{-126} \approx 1.18 \times 10^{-38}$. When you discover a *really* small number, it is often because you have misinterpreted an integer type as a floating-point type. The above example is contrived, but once you start messing with pointers and moving memory around, you are likely to see this type of result at some point, and now you will know why.

Here is one more example showing the consequences of adding disparate floating-point numbers. Try to guess how many iterations will occur before the following loop will exit:

```
float my_float =1.0;
while (my_float< 2.0) my_float += 1.0e-8;
```

This is, of course, a trick question. The result of adding 1.0 + 1.0e–8 will be rounded down to 1.0 as a float records approximately seven decimal places of accuracy, and therefore, the stored number will never increase, no matter how many iterations are executed. For this same reason, programmers are advised against using floating-point types as loop variables because rounding will affect the number of iterations. A similar phenomenon can occur when operations are performed in an indeterminate order, for example, a reduction in a parallel program. Adding small pieces first can generate larger pieces that would not be rounded down, but adding small pieces to much larger pieces may result in the small pieces being lost. If pieces are added in an indeterminate order, different runs can produce different results. Adding small numbers together first before adding them to larger numbers is something that needs to be handled carefully in parallel computing, where it is not always easy to specify the order in which parallel operations are performed.

2.7 Summary

As a computational scientist or engineer, knowing the pitfalls, implicit errors, and limits of working with machine numbers is fundamental to performing numerical calculations. Understanding the limits of representing decimals in machine code also has implications for statistics calculations and error analysis. For more information on numerical representation, the curious reader is directed to the IEEE 754 standard itself at http://standards.ieee.org/findstds/standard/754-2008.html.

3

Developing with Git and Github

Chris Ertel

CONTENTS

3.1 Introduction

When working on a codebase, whether it is a decades-old linear algebra package or just a small experiment script, it is important to be able to work without worrying about losing data.

One approach to preventing data loss is to save a copy of your files with a different name each time, usually appended with the date or revision number. Invariably, this ends up with a directory littered with files with names like "sample_data_v2," "preprocessor_final_rev_2," and "paper_draft_v2_2015_10_21_FINAL_3b." This can make finding the most authoritative version of a file difficult.

Another approach is to use e-mail to send copies of files being worked on back and forth between interested parties, grabbing the (hopefully) most recent version out of the message history and sharing that. This is error-prone, however, and makes it difficult to track which changes were made by which people—at least for things like source code files.

What we would like to have are tools that

- Allow us to share our work with others
- Prevent us from making accidental modifications to our work
- Allow us to compare different versions of our work to see changes

The best tools we have for this purpose are source control tools like cvs, svn, git, or any of several alternatives. We will be focusing on git.

Git is perhaps the most widely used tool in industry and open-source today, with a rich feature set and a large ecosystem built around it. Github is a member of this ecosystem and is currently the largest hosting provider for open-source and private projects.

Git is also useful because it is a distributed version control system (DVCS), which means that it can be used on your local machine without needing a central server. This is a different approach from, say, svn, which requires you to have an available connection to your server to get meaningful work done. This is particularly useful while, for example, traveling.

3.2 Basic Git

3.2.1 A Simple Overview of Git

It is helpful to have a basic mental model of git before trying to use it.

A file directory and its subfolders and files are checked into a *repository*. This is done by a `.git` folder with metadata being inserted into the top level of the directory tree.

The repository is a hidden `.git` folder that exists at the top level of the directory you want under source control. It contains the object store and other metadata about the project. The working directory is the normal folder you are doing your work in, where you change files and do work. The index is where files that are ready to be committed are staged. The index is where you batch changes to go into a single commit.

If you want an analogy, consider doing some work on your desk—albeit without the use of a computer! The papers on your desk are the files, and the desk itself is the working directory. The briefcase you put your papers into to take them home is the index; you only put the papers you want into it, leaving the other things on your desk undisturbed. Similarly, files not added to the index will not be included in a commit to the repository. Your home where all the papers are ultimately filed away is the repository.

There is an additional feature to git that sets it somewhat apart from version control systems like cvs or svn: the pervasive use of *branches*. A branch is a way of working on your code in an isolated context. When you are ready, you can merge branches together and thus add your progress to your main codebase. Other people can also maintain branches, and this makes it easy to share code with collaborators without ever having their changes (or yours!) unexpectedly break the code.

3.2.2 Creating a New Repository

At the start of a new project, you may like to create a new repository on your local machine. To do so, run this command:

```
~crertel$ git init test_repo
Initialized empty Git repository in /home/crertel/test_repo/.git/
~crertel$
```

This creates a new repository--test_repo--which can be used for work.

There is also a way of creating a "bare" repository using git init --bare. This will create a repository that is set up to be used by other users. It has no working directory, but instead only the metadata that supports git.

3.2.3 Cloning an Existing Repository

Alternatively, you can *clone* an existing repository. You will do this when starting work on a shared codebase, or *any* codebase stored in Github.

Cloning a repository creates a local copy of it on your machine. Your copy contains all of the history of the repository, as well as all active branches.

The command to clone a repository is of the form git clone <URL> and will produce output like so:

```
~crertel/projects$ git clone https://github.com/crertel/git-example.git
Cloning into 'git-example'...
remote: Counting objects: 3, done.
remote: Compressing objects: 100% (2/2), done.
remote: Total 3 (delta 0), reused 0 (delta 0), pack-reused 0
Unpacking objects: 100% (3/3), done.
~crertel/projects$
```

This would checkout a repository using HTTPS authentication, useful if you do not have SecureShell (SSH) keys setup. If those keys are set up, then it is better practice to use the git protocol:

```
crertel@rockbox:~$ git clone git@github.com:crertel/git-example.git
```

You will then have a subfolder in your current directory named after the repository you have checked out. In this case, the subfolder is named git-example.

3.2.4 Saving Work

To save your work and changes, you will need to *commit* them. This is the act of creating a batch of changes and marking it as a unit of work in your repository's history.

Once you have a file or files you would like to put into a commit, you can use git add <filename> to put them into the index. You can also use git add . to add every changed file in the current directory and below. You can then use git status to see that the files have been added.

```
~crertel/test_repo$ touch new_file.txt
~crertel/test_repo$ git add new_file.txt
~crertel/test_repo$ git status
```

```
# On branch master
## Initial commit
## Changes to be committed:
#   (use "git rm --cached <file>..." to unstage)
##    new file:   new_file.txt
#~crertel/test_repo$
```

To finally commit the changes in the index into the repository, you use the `git commit` command.

```
~crertel/test_repo$ git commit
[master (root-commit) 6831616] Add test file.
0 files changed
create mode 100644 new_file.txt
~crertel/test_repo$
```

When you run this command a text editor will appear to let you enter a commit message (not shown in the above session). Upon saving the temporary file that it will be editing, the commit will be registered with the repository.

If you need to remove a file that has already been committed, remove it normally from the working directory, run `git add -A` to update the index with its absence, and then commit your changes with `git commit`.

```
~crertel/test_repo$ rm b.txt
~crertel/test_repo$ git add . -A
~crertel/test_repo$ git commit
[master 2636345] Remove b.txt
0 files changed delete mode 100644 b.txt
~crertel/test_repo$
```

With this, you now know how to create a repository locally, how to update the files in that repository, and how to remove files from the repository. This is enough for basic work on your own machine, but git can do a lot more. Next, we will talk about sharing files.

3.2.5 Sharing Changes

If you cloned a repository from Github or some external source, you can also *push* your local changes back to that server. This is done with the `git push` command:

```
~crertel/scicomp-book-chapter$ git push origin
Username for 'https://github.com': crertel
Password for 'https://crertel@github.com':
Counting objects: 5, done.
Delta compression using up to 2 threads.
Compressing objects: 100% (3/3), done.
Writing objects: 100% (3/3), 532 bytes, done.
Total 3 (delta 1), reused 0 (delta 0)
To https://github.com/crertel/scicomp-book-chapter.git 5c876c3..22849a6
master -> master
~crertel/scicomp-book-chapter$
```

In order to fetch changes from the server (and thus get your collaborators' work), you run `git pull origin <branch name>`. That will bring any remote changes down into your local copy of the repository from the specified branch and merge them into your current branch.

An example run looks like this:

```
~crertel/scicomp-book-chapter$ git pull origin master
Username for 'https://github.com': crertel
Password for 'https://crertel@github.com':
From https://github.com/crertel/scicomp-book-chapter * branch
master      -> FETCH_HEADAlready up-to-date.
~crertel/scicomp-book-chapter$
```

This will make somewhat more sense when we cover branching in a moment.

3.3 Basic Github

Github (http://www.github.com) is where the majority of open-source development happens today. There are also other alternatives, like Bitbucket (https://bitbucket .org/).

Github provides repository hosting, easy user management, issue tracking, and a way for easily allowing people to clone your codebase and submit back patches.

3.3.1 Creating a Github Repository

Once you have created a Github account, you can create a repository from the new repository page (https://github.com/new), which can be accessed directly or from the "Create Repository" option of the "Create new..." menu in the upper-right-hand corner.

You will be prompted for

- A repository name
- A description of the repository
- Whether or not the repository will be public (viewable by anyone) or private (only to added collaborators)
- Whether to initialize it with a README file (containing the description of the repository)
- What gitignore file to use, in order to block out common artifacts
- What license file to include, in order to clarify how the code may be used and distributed

The page looks like this:

3.3.2 Adding People to a Repository

In order to give others read access (for private repositories) or write access (for all repositories), you need add their Github accounts to your repository. This can be done from the "Collaborators" section of the "Settings" menu.

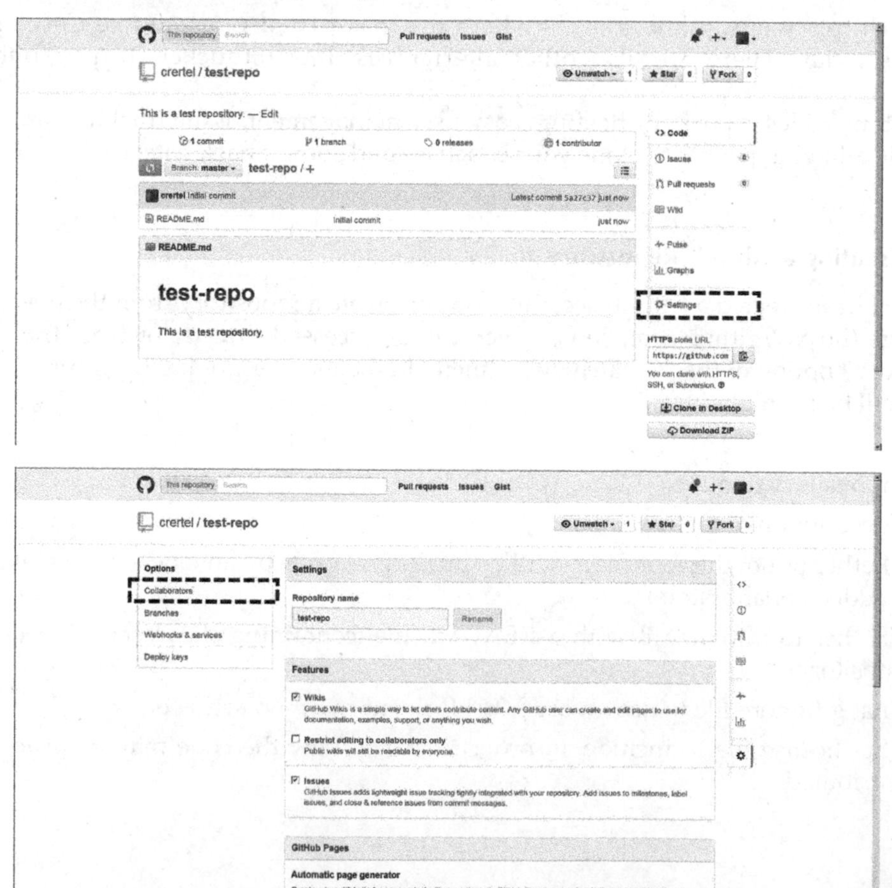

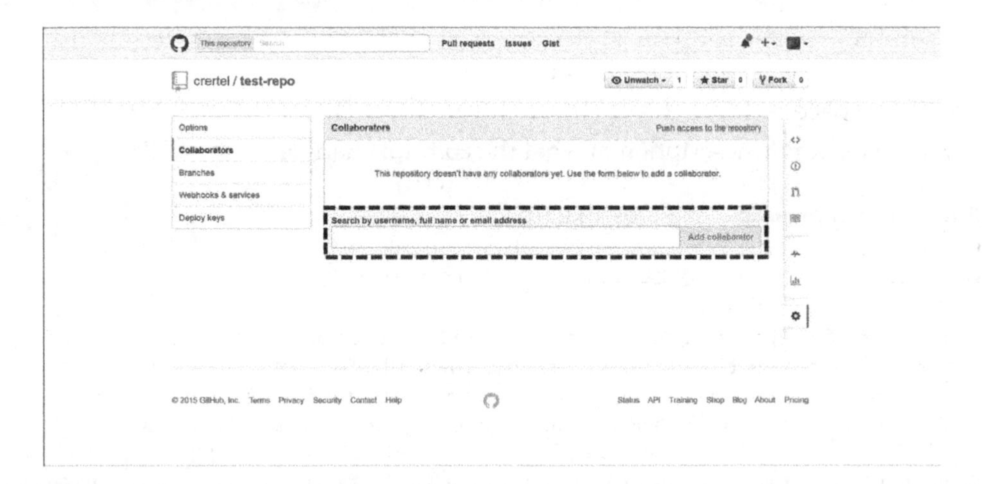

Once you have added collaborators, they will be able to read and write to the repository. This also means that they will be able to change the code and project history, so be careful. It may make sense to let collaborators *fork* your repository and submit code through *pull requests*, thereby ensuring your repository's integrity.

3.4 Intermediate Git

What we have covered thus far is basic tools usage. There are some more advanced workflows that git and Github support, as well as a few things that you can do to make your life easier.

3.4.1 Housekeeping

When working on a project, especially one with a lot of compiled artifacts (say, the .a and .out files in a C project), it is a good idea to leave those artifacts outside of source control. Git and similar tools are designed to be most effective with text files, and generally, binary artifacts only result in merge conflicts that have to be resolved at every commit. Debuggers and Integrated Development Environment (IDEs) may also create configuration and auto-save files (like .swp files from vim and .suo files from Visual Studio).

Git can be told to ignore these files by the addition of a `.gitignore` to the root directory the repository. This file is a simple file of the form:

```
a.out
bin/
```

which will cause files named a.out and anything in a directory named bin to be ignored. Each line is a different pattern to ignore, either a filename or a directory. Once ignored, files matching that pattern will not show up and cannot be added to the repository. Subdirectories may have their own `.gitignore` files as well if you do not want to have long path names in the root `.gitignore`.

It is important to make sure that you are writing useful commit messages. A good summary of how to write good commit messages is available at Torvalds (https://github.com/torvalds/subsurface/blob/master/README#L82-109) and yields commits of the following form:

- A subject line of fewer than 80 characters (to allow easy display in terminals) giving a brief explanation of the commit
- A blank space
- A more elaborate description of what the commit changes

Put together, a message may look like this:

```
Update the foo() routine to better bar bazzes.

The existing foo() routine sometimes fails to account for all bazzes.
This change updates the binary search used while barring to fix a case
where, with a bad comparator, a bar might not be found.
This should fix the problems we were seeing in issue #123.
```

It is useful to be able to review the history of changes made to a repository. This can be done with the `git log` command. Running this will provide the whole history of the repository run through the system pager. If you want, you can review only the last n entries in the log by invoking `git log -n`. An example of running this with $n = 3$ is shown below:

```
~crertel/scicomp-book-chapter$ git log -3
commit c908805f93bb245a797a9be67148b5ab1b035932
Author: Chris Ertel <chris@kedagital.com>
Date:   Mon Aug 3 13:03:19 2015 -0500
    Add section on commit messages.

commit 9785140aa9f5c7a87ac1d2cdeefbab659f4af78d
Author: Chris Ertel <chris@kedagital.com>
Date:   Mon Aug 3 02:04:50 2015 -0500
    Add section on .gitignore.

commit 978bbc9bdad609c6657cffff63f5b457fc131703
Author: Chris Ertel <chris@kedagital.com>
Date:   Mon Aug 3 01:15:46 2015 -0500
    Add "Basic Github" page.
~crertel/scicomp-book-chapter$
```

The `commit: 978bbc9bdad609c6657cffff63f5b457fc131703` line, also known as the *commit hash*, is the identifier for that particular snapshot in history of the repository. It can be abbreviated to the first few characters, which can be helpful for referring to a commit in commands and correspondence.

The `Author` line lets you see who made the change. The contact information sent out can be set by using:

```
~crertel/test_repo$ git config user.name "Chris Ertel"
~crertel/test_repo$
```

You can also set your global name, which is the default when you have not set a per-repository name:

```
~crertel/test_repo$ git config --global user.name "Chris Ertel"
~crertel/test_repo$
```

Your e-mail can be set in a similar fashion:

```
~crertel/test_repo$ git config user.email "chris@kedagital.com"
~crertel/test_repo$ git config --global user.email "chris@example.com"
~crertel/test_repo$
```

3.4.2 Fixing Mistakes

Even the best programmers make mistakes. The entire reason we have version control systems is to mitigate those mistakes. If you have changed a file and not committed it yet, you can use `git checkout` to reset the file to its previous version:

```
~crertel/test_repo$ echo "OOPS" >> new_file.txt ~crertel/test_repo$ git
status
# On branch master
# Changes not staged for commit:
#   (use "git add <file>..." to update what will be committed)
#   (use "git checkout -- <file>..." to discard changes in working
directory)
##   modified:   new_file.txt
#no changes added to commit (use "git add" and/or "git commit -a")
~crertel/test_repo$ git checkout new_file.txt
~crertel/test_repo$ git status
# On branch master
nothing to commit (working directory clean)
~crertel/test_repo$
```

This command can be done on an entire directory at once as well. It will replace the contents of the modified file with whatever was there at the time of the most recent commit, so be careful if you want to save the data.

If you have accidentally added a file you have not meant to, you can use `git rm --cached <filename>` to remove it from the index. If you want to remove it from the index and from the working directory, you can use `git rm -f <filename>`.

This technique works only if the file has not already been committed. In the case where it has been, you will need to *amend* the previous commit. This can be done by running the `git commit --amend` command.

In order to use this command, make the changes that you need to your repository, add them as usual, and then run the `git commit --amend` command. You will be given a chance to update the commit message, and then the most recent commit will be modified to reflect your new changes.

Note that doing this *should not* be done to commits that have been shared with your collaborators because it can create a repository history inconsistent with their local copies. Only do it for cleaning up local commits prior to sharing them.

More advanced cleanups can be handled with the `git rebase` command. Consult the git documentation on `git rebase`, as it is involved and outside the scope of this chapter.

3.4.3 Branching in Git

As mentioned earlier, branching is one of the most powerful features afforded to us by git.

In order to create a branch, we invoke the `git branch <branch name>` command, and then use the `git checkout <branch name>` command to make it active. This is shown below:

```
~crertel/test_repo$ git branch new_branch
~crertel/test_repo$ git checkout new_branch
Switched to branch 'new_branch'
~crertel/test_repo$
```

This can be done in one step by running `git checkout -b new_branch`.

```
~crertel/test_repo$ git checkout -b new_branch
Switched to a new branch 'new_branch'
~crertel/test_repo$
```

You can figure out what branch you are currently on by using `git status` or `git rev-parse --abbrev-ref HEAD`.

In order to see all the existing branches, run the `git branch` command:

```
~crertel/test_repo$ git branch
  master
  new_branch
* newer_branch
~crertel/test_repo$
```

An asterisk in that list denotes the current branch, in that case "`newer_branch`".

Any changes you make on the current branch will not affect other branches. If you attempt to checkout a different branch with unadded or uncommitted files, they will be brought along into the branch you are switching into. If the changed files are under source control, they will not be brought along and will remain with the branch they have been added to.

In Github, you can create a new branch using the branch dropdown. You select it, type in a branch name, and then can create that branch:

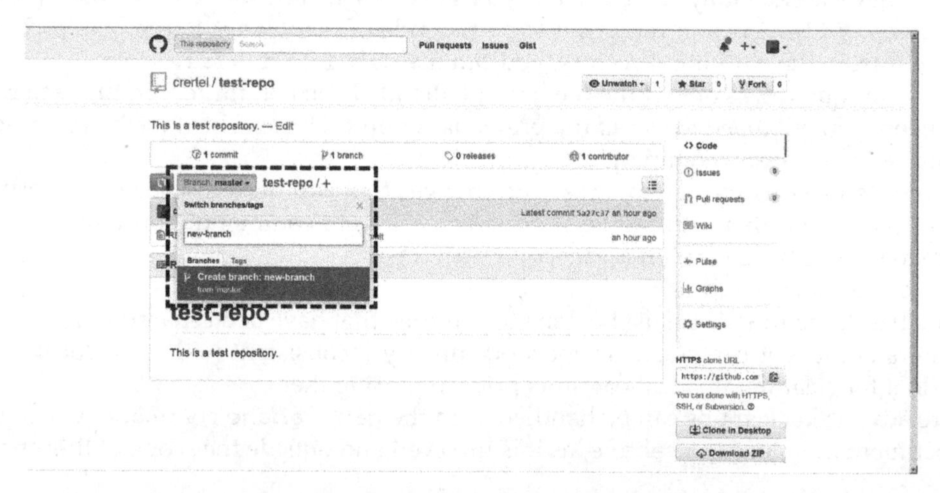

Once this is done, you will need to update your local repository by using the `git fetch origin` command and switch to that branch. An example of this being done is below:

```
~crertel/scicomp-book-chapter$ git fetch origin
Username for 'https://github.com': crertel
Password for 'https://crertel@github.com':
```

```
From https://github.com/crertel/scicomp-book-chapter * [new branch]
test        -> origin/test
~crertel/scicomp-book-chapter$ git checkout test M   chapter.md
Branch test set up to track remote branch test from origin.
Switched to a new branch 'test'
~crertel/scicomp-book-chapter$
```

3.4.4 Merging Branches

Once work on a branch is complete, you can merge its changes into another branch. This is generally an automatic process, done by switching to the branch you wish to merge into and running the `git merge` <from> command.

An example of this is as follows:

```
~crertel/test_repo$ git checkout -b to_merge_in
Switched to a new branch 'to_merge_in'
~crertel/test_repo$ touch test_text
~crertel/test_repo$ git add test_text
~crertel/test_repo$ git commit
[to_merge_in 7f04398] add test text
0 files changed create mode 100644 test_text
~crertel/test_repo$ git checkout master
 Switched to branch 'master'
~crertel/test_repo$ git merge to_merge_in
 Updating e25cc49..7f04398
Fast-forward
 0 files changed
 create mode 100644 test_text
~crertel/test_repo$
```

In this example, we have created and added a new branch to merge in `--to_merge_in`. Then, we created a new file. Then, we added and committed that file (writing a proper commit message in the editor). Last, we switched back to the `master` branch and then merge in the `to_merge_in` branch.

If you attempt to merge in a branch that has conflicts, for example, changes on the same lines in the same file, the merge will fail. It will look something like the following:

```
~crertel/test_repo$ git merge conflict
 Auto-merging test_text
CONFLICT (content): Merge conflict in test_text
Automatic merge failed; fix conflicts and then commit the result.
~crertel/test_repo$
```

In that case, you can run `git  status` to see which files are conflicted:

```
~crertel/test_repo$ git status
# On branch master
# Unmerged paths:
#   (use "git add/rm <file>..." as appropriate to mark resolution)
##   both modified:      test_text
#no changes added to commit (use "git add" and/or "git commit -a")
~crertel/test_repo$
```

The files will be annotated with the conflicting regions between the two branches:

```
<<<<<<< HEAD
This is version 2b.
=======
This is version 2a.
>>>>>>> conflict
```

HEAD is the version in the branch you are merging into. In order to resolve this conflict, decide which part (if any) of the conflicted regions you would like to keep, and remove the rest (including the annotations). Then, `git add` the updated file and commit your work.

3.5 Intermediate Github

3.5.1 Deleting Branches

You can remove branches from Github by using the branch management interface. All you have to do is pull up the interface, select the branch, and delete it.

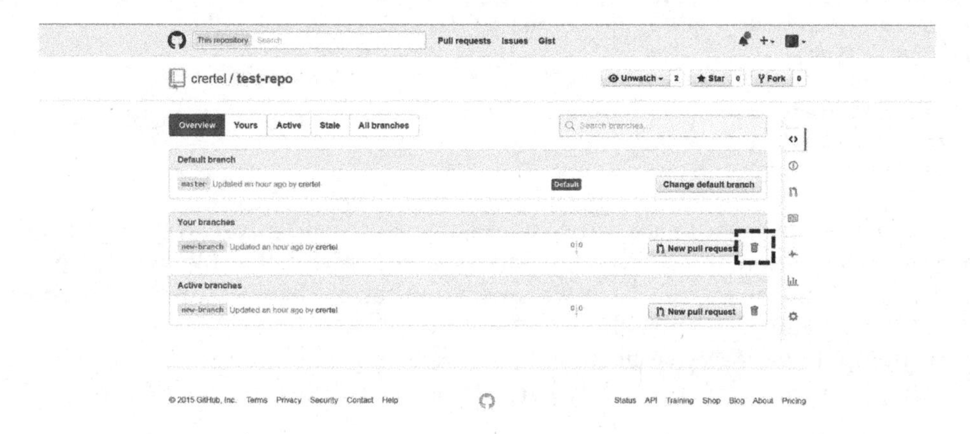

You can also remove a branch by pushing to it remotely:

```
~crertel/scicomp-book-chapter$ git push origin :test
Username for 'https://github.com': crertel
Password for 'https://crertel@github.com':
To https://github.com/crertel/scicomp-book-chapter.git - [deleted]
test
~crertel/scicomp-book-chapter$ git fetch origin -prune
Username for 'https://github.com': crertel
Password for 'https://crertel@github.com':
~crertel/scicomp-book-chapter$ git branch -D test
Deleted branch test (was fdcf1c5).
~crertel/scicomp-book-chapter$
```

Note that, first, we git push origin :<branch_name> to tell git to push the empty branch to the remote branch (deleting it). Next, we run git fetch origin −prune to make sure our local references are updated with the remote deletions. Finally, we run git branch -D test to completely remove all local traces of the branch we wanted to delete. Without all of these steps (in order), you will still have a copy of the branch on your local machine, which can be confusing.

3.5.2 Pull Requests

Github has a handy way of dealing with branching and merging, a mechanism called *pull requests.* A pull request is a collection of changes to a branch on Github, authored against another branch. Github provides a simple interface for seeing what changes will be merged and merging them. To open a pull request, go to the branches page and select the "New pull request" button.

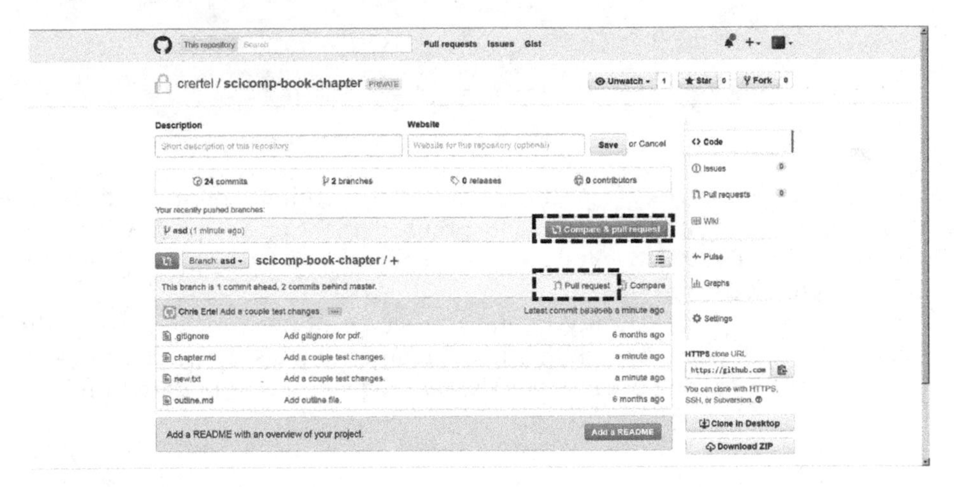

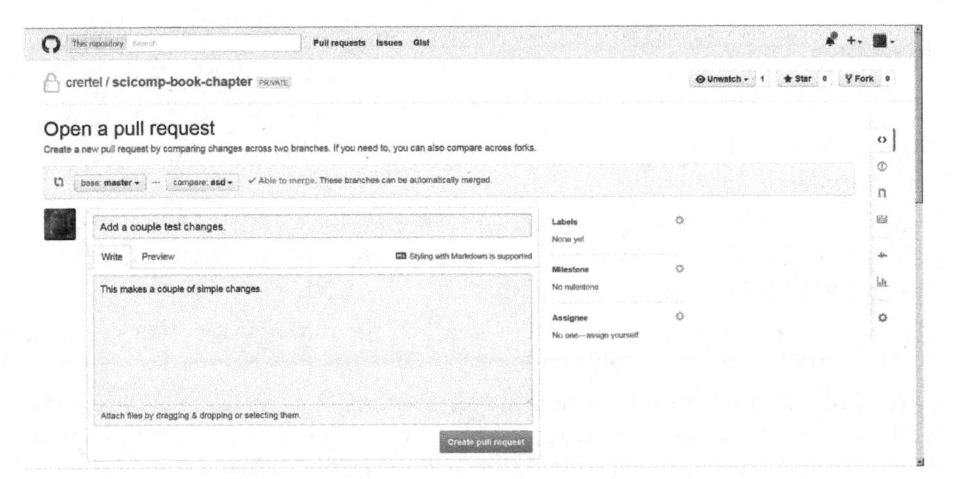

From here, you select the base branch to merge into and the compare branch to merge from. You may also give a description of the pull request, typically a simple explanation of what it changes and what issues (if any) it references.

To review a pull request, you will typically look at the summary of the changes in the form of changed files and commits. If you have changed many files, this can become difficult. In order to help avoid this, it is helpful to have small, focused pull requests. You can also comment on the request in its discussion thread, which can be useful for discussing changes.

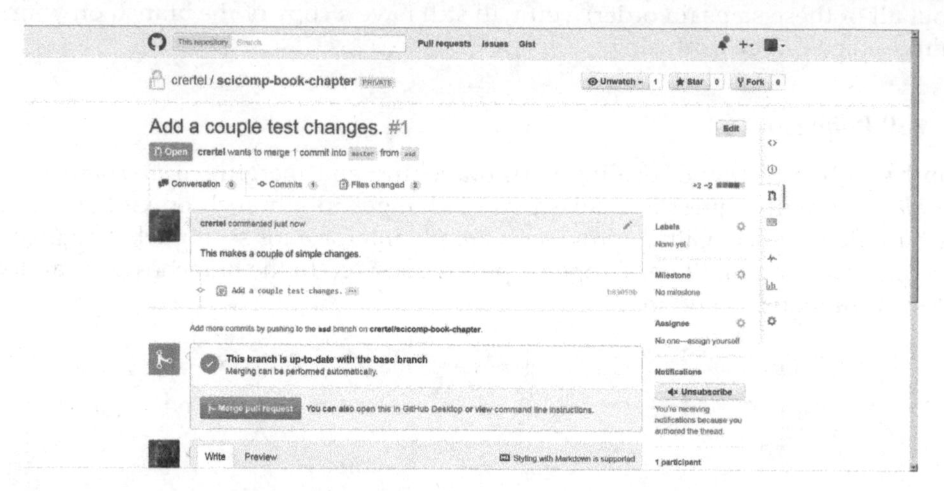

If you need to make changes to a pull request, simply continue committing on the compare branch. Once the commits are pushed to github, then the pull request will be updated. In order to merge a pull request, you need only to push the "Merge request" button at the bottom of the request summary page.

If the request cannot automatically be merged, you will need to first merge the base branch back into the compare branch, fixing the merge conflicts. Then, commit and push those changes back to the compare branch. The pull request will be updated, and then you can merge it.

3.6 Tips

- Commit frequently. It is cheap to do so and lets you backtrack easily if you change your approach.
- Write good commit messages—Follow the directions put out by Torvalds (https://github.com/torvalds/subsurface/blob/master/README#L82-109). This will make you and your collaborators' lives easier.
- Create one branch per new feature or bugfix. That makes it much easier to apply changes and to isolate unstable code.
- Never ever amend commits or rebase branches that have been or could have been seen outside your local system. This can cause weird errors for other people sharing your code, as their branch history may become out of sync.
- Don't check large binary files into the repository. Git is not optimized for that workload.

- Add build artifacts to the `.gitignore` file in the repository. If checked in, build artifacts only cause merge conflicts. Favor rebuilding locally over storing these objects.
- If using Github, be sure to include a simple `README.md` file in the root of the repository. This will allow the Github repo page to display helpful information about your project to whoever visits it. Consider including an overview of what the code does, any directions on how to report bugs or contribute patches, licensing information, and perhaps links to relevant academic work.
- Include a license of some sort in your repository if you are sharing your code. Doing so will help indicate to potential users your wishes on how the code is to be used.

Further Reading

Git, *The Git Reference Manual*, http://git-scm.com/doc.
Torvalds, Subsurface Project Contribution Guidelines, https://github.com/torvalds/subsurface/blob/master/README#L82-109.

4

Introduction to Bash Scripting

Erik E. Santiso

CONTENTS

In this chapter, we will cover one of the most useful tools for the implementation and automation of complex tasks in Unix-based systems: Bash scripting. Bash is one of the most common (if not the most common) shells used in these systems, particularly in Linux. As do all shells, Bash serves as an interface between the user and the kernel, and as such, it provides a language for handling system-level objects (such as jobs, processes, and files) with ease. Bash implements a "read–evaluate–print loop" (REPL) environment, which reads user input, evaluates it, and prints the result. This type of environment enables rapid development and debugging, compared to classical compiled languages. Even though there are scripting languages that are easier to learn and have a large code base in the form of libraries (such as Python), Bash is often used to write "master scripts" to combine code from different sources and manage files and processes, as those tasks are much easier to implement in Bash. Bash scripts are also highly portable, as nearly every Unix-like system will include the Bash shell. More details beyond what is covered in this chapter can be found in the references at the end of the chapter [1,2]. The examples in this chapter were tested with GNU Bash version 3.2.48 (the version of Bash running in a system can be found by running `bash –version`).

4.1 Introduction: Building Your First Bash Script

Scripting is one of the most powerful features of Unix-based operating systems. It enables the automation of repetitive tasks with ease and allows for complex tasks requiring the interaction of several different programs to be carried out smoothly.

Scripts are sequences of instructions, or programs, in one of several different *scripting languages*. Unlike programs written in a compiled language such as Fortran or C++, scripts are read and interpreted at runtime, whereas compiled languages are first translated into machine code by a compiler and then run. Usually, scripting languages have easier and more flexible syntax than compiled language does and are intended to automate high-level tasks. However, because they are interpreted at runtime, they are not optimal for tasks where speed is important (e.g., calculation-intensive simulations).

There are many scripting languages, and we already mentioned a few in Chapter 1 (such as awk and sed). Other commonly used languages are Python, which is discussed in a separate chapter, Perl, tcl, Ruby, Javascript, and PHP (the last two are ubiquitous in web-based applications). In this particular chapter, we will focus on Bash scripting, which utilizes commands from the Bash shell described in Chapter 1.

Our first example will follow the long-standing tradition of getting the computer to say "Hello, world." In a terminal running Bash, use a text editor to create a file called "myScript" with the following lines:

```
#!/bin/bash
# Print "Hello, world!" and exit
echo Hello, world!
exit 0
```

The first line is a "hashbang" or "shebang" tag. The purpose of this line is to define what interpreter will be used to run the remaining lines on the script when the script is run as a program. In this case, it indicates that the script should be interpreted by the Bash shell (which is assumed to be located at /bin/bash). You can use any program that is able to interpret text commands in the shebang, for example, if you are writing a Python script, you could use `#!/usr/bin/python`. If you are not sure of the location of the interpreter in your filesystem, you can use the `which` command to find it. For example, `which python` will return the location of the python command (if it is located within a directory within your `$PATH` environmental variable):

```
$ which python
/usr/bin/python
```

The second line in the script is a comment. Anything in a line following the hash ("#") character is ignored by Bash and can be used to explain what the script is doing. Commenting is an extremely important programming practice, not only if you are sharing your code with someone else, but also for code reusability: you may open the same file 3 months later only to find you do not remember how it works. The hash character can be anywhere within the line, for example, the script above could also be commented this way:

```
#!/bin/bash
# Print "Hello, world!" and exit
echo Hello, world!      # Say "Hello, world!"
exit 0                  # Exit
```

The third line in the script uses the `echo` command to print the string "Hello, world!" to standard output, and the fourth line tells Bash to exit with a return code of 0. This is optional, but it is good programming practice as it allows the calling program/user to know whether a problem occurred during execution (usually the return code "0" is used to mean that the code executed without errors). The return code can be obtained after running the script by using the $? variable as shown in the next code snippet.

After creating the script, we need to make it executable. This can be done with the chmod command, discussed in Chapter 1. For example, you can run `chmod a+x myScript` to make the script executable by all users or `chmod u+x myScript` to make it executable only by yourself. In this case, it doesn't make a big difference, but depending on the purpose of the script, you might want to restrict other users from executing the script for security reasons.

Once you have made the script executable, you can run by using `./myScript`. If you want to see the return code of the script, you can run `echo $?` immediately after executing the script:

```
$ ./myScript
Hello, world!
$ echo $?
0
```

The './' is needed because, in general, the current directory (i.e., './') would not be included within the `$PATH` environmental variable. The './' tells Bash explicitly where the script file to execute is.

4.2 Variables and Substitution

Like most other programming languages, Bash can use variables to store data. Bash is a *weakly typed* language, as variable types do not need to be explicitly defined. All Bash variables are strings of text by default, but depending on context, they may be interpreted as integer numbers. Bash does not allow operations with floating-point numbers directly, but floating-point operations can be handled through pipes to commands that allow them, such as `bc` or `awk`. Variables can, however, be declared as a particular type or given special attributes using the `declare` command. In particular, a variable can be declared as integer using `declare -i`, as a read-only variable using `declare -r`, and as an array using `declare -a`. Declaring a variable allows using simpler syntax for operations involving the variable. For example, if a variable is declared as integer, arithmetic expressions assigned to it get directly evaluated without using arithmetic expansion. You can see the current attributes of a given variable using `declare -p`:

```
$ declare -i MY_NUMBER
$ declare -r MY_NUMBER
$ declare -p MY_NUMBER
declare -ir MY_NUMBER=""
```

To define the value of a variable in Bash, simply use the equal sign, "=", with no spaces before or after. For example, to define a variable called MY_AGE with a value of "20", you would write `MY_AGE=20`. Note that using the variable name directly does not replace the name with its value in Bash. For example, running `echo MY_AGE` after the definition above would return "MY_AGE" instead of 20. To access the data stored in the variable, you must precede the variable name with a dollar sign, "$." In out example, `echo $MY_AGE` would output "20."

The example above illustrates the use of *parameter expansion*. When a dollar sign, "$", is placed before a variable name, it replaces the name with the data stored in it. More complex parameter expansion can be done using the syntax ${*expression*}, where the braces enclose an expression to be expanded. If the braces only enclose a variable name, the result is the same as that without the braces; that is, the variable is replaced with its contents. Some other uses of parameter expansion are described in Ref. [3]:

- The dash character, "-", can be used within a parameter substitution to test whether a variable name has been used or not. The variable to be tested should be at the left of the dash, and the parameter to evaluate, if it does not exist, should be at the right. For example, if we have set the value of MY_NAME to "Erik", the expression `${MY_NAME-Batman}` will evaluate to "Erik", whereas if the variable MY_NAME has not been defined, the same expression will evaluate to "Batman":

```
$ echo ${MY_NAME-Batman}
Batman
$ MY_NAME="Erik"
$ echo ${MY_NAME-Batman}
Erik
```

- The colon character, ":", can be combined with "-" to test whether the variable is null in addition to its existence. For example, if MY_AGE is set to a null string, the expression ${MY_AGE-Undefined} would return the null string, but ${MY_AGE:-Undefined} would return UNDEFINED:

```
$ MY_AGE=""
$ echo ${MY_AGE-Undefined}

$ echo ${MY_AGE:-Undefined}
UNDEFINED
```

- The characters ":=" can be used to do conditional assignment. In the expression ${YOUNG:=10}, if the variable YOUNG has not been defined (or contains a null string), the value 10 will be stored in it and the expression will evaluate to 10. If the variable has been defined, it will stay unchanged and the expression will evaluate to its current contents.
- The exclamation mark can be used to evaluate a variable name stored within a variable. For example, if MY_AGE is set to 20, and MY_VAR is set to MY_AGE, the expression $MY_VAR will evaluate to MY_AGE, but ${!MY_VAR} will evaluate to 20.
- Substrings can be extracted from a variable using the notation "${*var*:*offset*:*length*}". In this case, *var* is the variable containing the string, *offset* defines the starting point, and *length* is the number of characters to expand. If the length is omitted, the string is expanded until its end. For example, if we set the variable MY_STRING to the string "This is a long string", then ${MY_STRING:10:4} will evaluate to "long", and ${MY_STRING:14} will evaluate to "string". An ordinal sign, "#", in front of the name of the string will return the length of the string; in our example, ${#MY_STRING} would evaluate to 21, as the string contains 21 characters.

There are more complex uses of parameter expansion, examples of these can be found in Ref. [1]. Note that, besides these, some the commands described in Section 1.5 (such as `printf`, `cut`, `head`, `tail`, and `sort`) can also be used to extract information from variables.

The dollar sign "$" has two other common uses in Bash besides parameter expansion [1]. One is *arithmetic expansion*. The syntax for arithmetic expansion is "$((*expression*))", where the double parentheses contain the arithmetic expression to calculate. For example, if the variable MY_AGE contains the value 20, then $((3*MY_AGE+2)) will evaluate to 62. Note that assigning an arithmetic expression directly to a variable that has not been declared as an integer will store the expression, not the result. For example, if MY_AGE has not been declared as an integer, assigning MY_AGE=3*20+1 will result in MY_AGE containing the string "3*20+1", not the value 61. Instead, MY_AGE=$((3*20+1)) will store the value 61 in MY_AGE. If, however, the variable has been declared as an integer using `declare -i MY_AGE`, assigning MY_AGE=3*20+1 will store the result 61 into MY_AGE. An alternative syntax for arithmetic calculations is using the `let` command, which enables assigning the result of an arithmetic operation to a variable using a simpler syntax. For example, the result above could also be achieved by running `let MY_AGE=3*20+1`.

The other use of the dollar sign is *command expansion*. The syntax in this case is "$(*command*)", where *command* is a shell command to be executed. An alternate syntax is using backquotes, "`" to surround the command; that is, `` `command` `` will also evaluate to the standard output of the command:

```
$ echo $(ls -d D*)
Desktop Documents Downloads
$ echo `ls -d A*`
Applications Archive
```

This is particularly useful when writing Bash scripts, as it allows us to manipulate and interpret the result of shell commands. Some additional examples of this usage are given in later sections.

Finally, variables in Bash can be used to store arrays. There are a few ways to define an array variable in Bash. One is to declare it as an array using `declare -a`. Another way is to directly assign a value to an element of the array enclosed within square brackets. For example, assigning MY_ARRAY[1]=25 will define MY_ARRAY as an array variable and store the value 25 in it as the second element (array indices count from zero). Finally, an entire array can be assigned by listing its elements within parentheses; for example, if we define the variable MY_ARRAY=(a b c d e), then `${MY_ARRAY[2]}` will evaluate to "c". If the index is omitted, an array variable evaluates to the first element of the array; so `$MY_ARRAY` in our example would evaluate to "a". Using the at symbol "@" or the asterisk, "*", as the argument will return all the arguments of the array, but if used with a numeral sign, "#", before the name of the array, it will return the number of elements. For example, `${MY_ARRAY[@]}` in our example will return "a b c d e", and `${#MY_ARRAY[@]}` will return 5.

As an example, the script below defines an array of names and prints to standard output the number of names stored in the array, the first and last name stored, and the names sorted in alphabetical order:

```
#!/bin/bash

# Define the array
NAMES="Doc" "Grumpy" "Happy" "Sleepy" "Bashful" "Sneezy" "Dopey")
```

```
# Store the length
LENGTH=${#NAMES[@]}

# Write the length, and the first and last elements
echo There are $LENGTH names in the array
echo The first name stored is ${NAMES[0]}
echo The last name stored is ${NAMES[$LENGTH-1]}

# Write the sorted names
echo These are all the names sorted alphabetically:
echo ${NAMES[*]} | xargs -n 1 | sort
```

Running this script will produce the following output:

```
There are 7 names in the array
The first name stored is Doc
The last name stored is Dopey
These are all the names sorted alphabetically:
Bashful
Doc
Dopey
Grumpy
Happy
Sleepy
Sneezy
```

4.3 Input and Output

Often, a script should be able to handle information defined at runtime. Information can be passed to a script via environmental variables (discussed in Chapter 1) or via command-line arguments. Command-line arguments are accessible from a Bash script via parameter substitution: the syntax is $*number*, with *number* being the index of the argument. The number 0 is used for the command itself, and the arguments start with 1. If there are 10 or more arguments, the number must be enclosed in braces. For example, the 12th argument would be ${12}. Alternatively, the `shift` command can be used to reassign the parameters (for example, shift 3 would assign $4 to $1, $5 to $2, etc.). Note that $0 is not reassigned by shift.

As an example, when running the following command:

```
myscript arg1 arg2 arg3 arg4
```

within the script, $0 will evaluate to "myscript", $1 to "arg1", $2 to "arg2, and so on. If the script contains the line "shift 2", after that line, $0 will still evaluate to "myscript", $1 will evaluate to "arg3", and $2 will evaluate to "arg4" ($3 and $4 will not be null).

A few other variables within a script contain information about the command-line arguments. In particular, $# returns the number of arguments, and $@ and $* return all the arguments: $@ returns them as separate words, and $*, all as a single word ($@ is usually better). For example, the following script will return the number of arguments, a list of the arguments on a single line, and then a list with one argument per line:

```
#!/bin/bash
# Write the number of arguments
echo I got $# arguments
# Write the arguments as a single line
echo $*
# Write the arguments one line at a time
echo Line by line:
echo $@ | xargs -n 1
```

For example, if the script above is named `checkArgs`, we could use it as

```
$ ./checkArgs Superior Michigan Huron Erie Ontario
I got 5 arguments
Superior Michigan Huron Erie Ontario
Line by line:
Superior
Michigan
Huron
Erie
Ontario
```

Note that `echo $* | xargs -n 1` would not be different from `echo $*`, since `$*` returns the arguments as a single object.

Scripts can write output to one or multiple files. This can be achieved by directly using redirection (with the "<", ">", and ">>" syntax; see Chapter 1, Section 1.2) or by referring to files using *file descriptors*. File descriptors are numbers associated with files. To refer to the corresponding file, the number is preceded by an ampersand, "&". Scripts, just like any other Unix command, have an input stream (called stdin) and two output streams (called stdout and stderr). By default, the file descriptors for stdin, stdout, and stderr are 0, 1 and 2, respectively. Additional files should use numbers greater than 2.

Opening a file using file descriptors can be done by running one of the following commands:

```
exec descriptor>filename    # open file for writing only
exec descriptor<filename    # open file for reading only
exec descriptor<>filename   # open file for reading and writing
```

For example, `exec 3>myfile` will open the file "myfile" for writing and assign the file descriptor "3" to it. If the file does not exist, it will be created. To close a file, the syntax is *descriptor*>&-. For example, `exec 3>&-` will close the file with the descriptor number 3. Reading from a file can be done using the "<" syntax or the `read` command. In particular, `read varname < filename` will read a line from the given file and store it in the variable *varname*. Alternatively, `read -n numchars varname < filename` will read the given number of characters instead of a full line, and `read -a arrayname < filename` will read a line and store each word separately as an element of the array *arrayname*. The read command will exit with the code 0 if reading was successful, 1 otherwise (this is useful when reading a file within a loop, discussed in the next section).

Output to one stream or file can be redirected to another. For example, running `exec 2>&1` will cause standard error to be written to standard output. If the first number is omitted, it is assumed to be 1 (stdout) by default.

As an example, the following script will write "Hello World" to file "myfile" and will read the words back from the file into an array. It will then write each element of the array into a separate line.

```
#!/bin/bash
# Write "Hello World" to "myfile"
echo Hello World > myfile
# Use "myfile" for reading
exec 3<myfile
# Read the first line of "myfile" into an array
read -a myarray <&3
# Close the file
exec 3>&-
# Write the arguments one line at a time
echo ${myarray[@]} | xargs -n 1
```

Running this script will create the file "myfile" containing the words "Hello World" and return:

```
$ ./myScript
Hello
world
```

In this example, the fifth line assigns the descriptor number 3 to the file "myfile". There must not be a space between the number and the descriptor, or Bash will interpret "3" as a command to run. However, there can be a space before the file name, which will be ignored (i.e., `exec 3< myfile` would be equivalent). The seventh line uses "read" to construct an array called "myarray" read from the file with the descriptor number 3 (i.e., "myfile"). It is important to have no space between "<" and "&". A space before the "3" would still work, but it is not common usage. Finally, the ninth line closes the file. Again, there should be no space between the "3", ">", and "&", but there can be a space before the dash (although it is not common usage).

4.4 Flow Control

The ability to use loops and conditional execution is key to any programming language, and Bash is no exception. This section will cover the most common commands used for flow control in Bash.

Conditional execution in Bash is usually implemented using the `if` or the `case` commands. In the first case, a condition is tested and execution branches based on whether the condition is true or false. In the second case, the execution branches based on multiple conditions.

The syntax for the `if` command is

```
if condition
then
  commands to execute if condition is true
else
  commands to execute if condition is false
fi
```

The condition to test is, in general, any Bash command. If the exit code of the command is zero, the condition is assumed to be true, whereas a nonzero exit code is interpreted as false (note that this is the opposite of many other programming languages). Often, the command used to test conditions is the Bash Built-in `test` command. This command enables a wide variety of tests on files, strings, or integers. The test command returns a zero as its exit code if the condition is true, and a one if it is not. Below is a list of some common uses of `test` (for more, consult the `man` page for `test`):

```
test -f filename     # Tests that a file exists
test -d dirname      # Tests that a directory exists
test string          # Tests that a string is the null string
test -z string       # Tests whether the length of a string is 0
test -n string       # Tests whether the length of a string is not 0
test str1 = str2     # Tests whether two strings are equal
test str1 != str2    # Tests whether two strings are different
test num1 -eq num2   # Tests whether two integers are equal
test num1 -ne num2   # Tests whether two integers are different
test num1 -gt num2   # Tests for num1 greater than num2
test num1 -ge num2   # Tests for num1 greater than or equal to num2
test num1 -lt num2   # Tests for num1 less than num2
test num1 -le num2   # Tests for num1 less than or equal to num2
```

Multiple conditions can be tested by using the –a (logical and) and –o (logical or) options. For example, `test $num -gt 0 -a $num -lt 10` will test that the variable *num* contains a value between 1 and 9. Also, the logical not operator is available as an exclamation mark, "!". For example, `test ! $num -ge 0` will test that the variable *num* contains a negative number.

A shortcut for `test` is using *test brackets.* Instead of explicitly typing `test`, the arguments to `test` can be enclosed in square brackets, leaving a space between both brackets and the arguments. For example, the following two commands are equivalent:

```
if test -e filename
if [ -e filename ]
```

One advantage of the test bracket notation is that it looks closer to the syntax for conditions in other programming languages. An alternate notation that is widely used is the double bracket, or *extended test* notation. The double bracket enables comparison using "<", ">", "==", and "!=" and using the or and and operators "||" and "&&", respectively, with the same syntax as the C language. For example, the following will test whether a variable contains a number greater than 5 or less than 2:

```
if [[ $var > 5 || $var < 2 ]]
```

Additionally, the double brackets allow the use of the "=~" operator, which tests whether a variable matches an extended regular expression (see Chapter 1). For example, the following tests whether a variable contains an integer number:

```
if [[ $var =~ ^[+-]?[0-9]+$ ]]
```

Multiple `if` commands can be nested using `elif`, which is short for "else if":

```
if condition1
then
  commands to execute if condition1 is true
elif condition2
then
  commands to execute if condition2 is true
else
  commands to execute if neither is true
fi
```

For example, the following would test whether a variable contains an even number, an odd multiple of three, or neither:

```
if [ $(( var%2 )) -eq 0 ]
then
  echo $var is even
elif [ $(( var%3 )) -eq 0 ]
then
  echo $var is an odd multiple of three
else
  echo $var is neither even nor a multiple of three
fi
```

An alternative way to nest multiple comparisons is the `case` command. The syntax for `case` is

```
case expression in
pattern1)
  commands to execute if expression matches pattern1
  ;;
pattern2)
  commands to execute if expression matches pattern2
  ;;
(...)
*)
  commands to execute if expression does not match any pattern
  ;;
esac
```

The patterns in the `case` command are *shell patterns*, which have a similar but much simpler syntax than regular expressions. In particular, they can include braces (which will match any of the characters within) and use the "*" and "?" special characters to match any string or any single character, respectively. Multiple options in a pattern can be separated by "|", which acts as an `or` operator. For example, the following code will identify a string as the name of a planet or dwarf planet in the inner or outer Solar System, regardless of whether the first letter is capitalized:

```
case $string in
[mM]ercury|[vV]enus|[eE]arth|[mMars])
  echo $string is an inner solar system planet
  ;;
[jJ]upiter|[sS]aturn|[uU]ranus|[nN]eptune)
  echo $string is an outer solar system planet
  ;;
[cC]eres)
  echo $string is an inner solar system dwarf planet
  ;;
[pP]luto|[eE]ris|[hH]aumea|[mM]akemake)
  echo $string is an outer solar system dwarf planet
  ;;
*)
  echo I do not know a planet called $string
  ;;
esac
```

In addition to conditional execution, an important programming tool is the ability to use loops. There are three common commands to define loops in Bash: `for`, `while`, and `until`. These commands are similar to those found in other programming languages. The `for` command executes a series of instructions repeatedly while assigning a variable values from a list. The syntax for this kind of loop is

```
for var in list
do
  commands
done
```

In this case, the variable *var* is successively assigned each of the values from *list*, and the commands between "do" and "done" are executed sequentially, with *var* containing each of the values. The list can be defined by explicitly listing the elements, by using command substitution, or by using a *brace expansion*, explained below. In the first case, the list is defined by enumerating the elements, separated by spaces, for example:

```
for dwarf in Doc Grumpy Happy Sleepy Bashful Sneezy Dopey
do
  echo Hi Ho, off to work $dwarf goes
done
```

Alternatively, the list can be the output of a command, using command substitution with either the "$()" or the backquote syntax as in this example. The following will rename all files and directories in the current directory by appending the suffix "_old" to them:

```
for file in `ls -d *`
do
  mv $file ${file}_old
done
```

Combining this syntax with the seq command allows for looping over a set of numeric values, for example:

```
for num in `seq 4 -1 1`
do
  echo $num little monkeys jumping on the bed
  echo one fell off and bumped his head
done
```

This would produce the following output:

```
4 little monkeys jumping on the bed
one fell off and bumped his head
3 little monkeys jumping on the bed
one fell off and bumped his head
2 little monkeys jumping on the bed
one fell off and bumped his head
1 little monkeys jumping on the bed
one fell off and bumped his head
```

The third common way to define the list of values is by using brace expansion. A list of elements enclosed in braces is automatically expanded by Bash to generate a list. The braces may enclose a list of arguments separated by commas, for example:

```
echo "Old MacDonald had a "{cow,chicken,dog}.
```

will produce the output:

```
Old MacDonald had a cow. Old MacDonald had a chicken. Old MacDonald had a
dog.
```

A numeric or alphabetic sequence can be defined using "..". For example, {1..5} will expand to the list "1 2 3 4 5", and {a..f} will expand to the list "a b c d e f". Note that mixing numbers and characters will not work. Newer versions of Bash allow for generating zero-padded numbers automatically by adding leading zeros to one of the numbers in the range, for example, {01..10}, but this is not supported in all computers. Additionally, newer versions of Bash allow to define an increment using the syntax {*first..last..increment*}, but this is also not supported in old versions. Note that both of these behaviors can be achieved using `seq` in a more portable manner.

When multiple brace expansions are mixed, all possible combinations are generated. For example, {a..e}{1..4} will expand to "a1 a2 a3 a4 b1 b2 b3 b4 c1 c2 c3 c4 d1 d2 d3 d4 e1 e2 e3 e4". If this is not what is intended (for example, you instead want to generate the list "a b c d e 1 2 3 4"), the braces should be nested and separated by commas, for example, {{a..e},{1..4}}.

The second common way to define loops in Bash is using the `while` and `until` commands. In the first case, the commands within the loop are executed while a condition is true. In the second, they are executed until the condition becomes true (i.e., while it is false) The syntax is as follows:

```
while/until condition
do
  commands
done
```

The condition to be tested follows the same format as conditions used for the `if` command. As an example, the following code is equivalent to the "monkeys" code in the previous example:

```
num=4
while [ $num -gt 0 ]
do
  echo $num little monkeys jumping on the bed
  echo one fell off and bumped his head
  num = $(( num-1 ))
  done
```

A useful idiom to read files is `while read` *`line`*. As explained on the previous section, the `read` command will read files line by line. If a line is read, its exit status will be zero, whereas the exit status will be one if a line was not read (for example, the end of the file was reached). Therefore, `while read` *`line`* will loop through a file until the end of the file, storing each line on the variable *line*. As an example, the following will print each line of the file named "myfile" preceded by its line number:

```
numLines=0
while read line
do
  let "numLines += 1"
  echo Line ${numLines}: $line
done < myfile
```

In addition to the above, there are three commands in Bash that are useful for flow control: `continue`, `break`, and `exit`. The `continue` command will skip the rest of a loop and continue with the following step. In the case of a `for` loop, the iterating variable will take its next value, and in the case of `while`/`until`, the condition will be checked again. When used in a nested loop, `continue` skips to the next iteration of the innermost loop. It is possible to skip to the next iteration of an outer loop by using `continue` *`number`*. For example, `continue 2` will skip to the next iteration of the loop containing the innermost one.

The `break` command is similar to `continue`, but it immediately exits the loop instead of skipping to the next iteration. As with `continue`, `break` will exit the innermost loop in a set of nested loops, but `break` *`number`* can be used to break out of the enclosing loop(s).

Finally, the `exit` command immediately ends the execution of a script and returns an exit code. If no exit code is given, it returns the exit code of the last command executed. Usually, `exit 0` (exit code 0) is used to denote successful execution, and other exit codes can be used to denote various errors.

There is one more mechanism for flow control in Bash, associated with signal handling. As mentioned in Chapter 1, signals are used by processes to communicate. If desired, it is possible to "capture" signals and execute code in response using the `trap` command. The syntax is

```
trap command signals
```

where *command* is the command to be executed when one of the signals listed is caught (recall that SIGKILL cannot be handled). This is useful to allow for "cleaning up" temporary files or variables created by the script. For example, the following will catch a SIGINT signal (sent when the user hits CTRL-C), a SIGHUP signal (sent when the invoking shell exits), or a SIGTERM signal and clean up the temporary file *temp*:

```
trap "rm -f temp; exit" SIGINT SIGHUP SIGTERM
```

More complex cleanup can be implemented by defining a cleanup function that is then called via trap (see Section 4.5).

As a final example, the following code will continuously ask the user to choose from a list of animals and print the corresponding animal sound to the screen. The `break` command is used to exit the `while` loop (otherwise the loop will never exit, as the condition is always true), and the program returns an exit code of zero upon successful execution.

```
while true
do
echo "Select an animal (0 to exit):"
  echo "1) Cow"
  echo "2) Donkey"
  echo "3) Dog"
  echo "4) Cat"
  read choice
  case $choice in
    1) echo Moo
    ;;
    2) echo Hee-haw
    ;;
    3) echo Bow wow
    ;;
    4) echo Meow
    ;;
    0) break
    ;;
    *) echo "Please select a valid option"
    ;;
  esac
done
echo "Good bye!"
exit 0
```

4.5 Functions

As in most other languages, Bash allows for defining functions. Using functions is a good mechanism to ensure the maintainability and readability of complex scripts. There are two ways to define functions in Bash. One is to use the `function` keyword:

```
function name
{ commands }
```

Where *name* is the name of the function. The braces enclose the commands to execute when the function is called. The braces can be in different lines. Alternatively, a function can be defined by simply following the function name with parentheses:

```
name()
{ commands }
```

Arguments can be passed to functions using the same syntax for passing arguments to scripts (see Section 4.3): `$1` represents the first argument, `$2` the second, and so on. For example, the following function will add two integers and print the result, or print an error if one of the arguments is not an integer:

```
function add
{
if [[ ($1 =~ ^[+-]?[0-9]+$)
     && ($2 =~ ^[+-]?[0-9]+$) ]]
  then
    echo $(( $1+$2 ))
  else
    echo Error: Arguments must be integers
  fi
}
```

Functions can be defined in separate files and included in a script using the `source` command. This eases the task of organizing a complex script and enhances maintainability.

In practice, it is common to encounter utilities to carry out different tasks written in different languages, or compiled as binaries. In such situations, it is very common for a Bash script to control the workflow. External code can be encapsulated in functions, and code written in `sed`, `awk`, or `python`, can be directly embedded in a Bash script to carry out complex tasks. This is probably the most common use of Bash scripts, other than automating very simple tasks.

References

1. C. Newham, and B. Rosenblatt, *Learning the Bash Shell*, 3rd Ed., O'Reilly, Sebastopol, CA, 2005.
2. M. Cooper, *Advanced Bash-Scripting Guide*, http://www.tldp.org/LDP/abs/html/.
3. Bash Reference Manual, https://www.gnu.org/software/bash/manual/bash.html.

Further Reading

Albing, C., J.P. Vossen, and C. Newham, *Bash Cookbook*, O'Reilly, Sebastopol, CA, 2007.

5

Debugging with gdb

Frank T. Willmore

CONTENTS

5.1 Print Statements and Macros

A great many computational scientists and engineers, myself included, have made their way through graduate school and beyond without the aid of a debugger. This is unfortunate, as once I finally spent a few hours learning the basics of `gdb`, I realized just how much time I might have saved. Yes, it's true, one can use print statements, mind-bending macros, and endless recompiles to achieve the same results. And while one can also use a rock to successfully hammer a nail, I personally prefer to use a nail gun. There are good tools available. Take the hit and learn to use a debugger.

5.2 Sample Code

Every great programming tutorial needs a great sample code. This will have to do. This is an example of a test program that might be used to generate an array of integers, with their values initialized to their index, that is, `array[n] = n`. It takes one command line argument, which will be the requested length of the array. The code also prints out the contents of the array that has been constructed.

This program contains two issues. The first will manifest itself on line 12 of Figure 5.1. The `atoi()` function, which does not provide error checking, is used to convert a string of

```
 1  //buggy.c
 2
 3  #include <stdio.h>
 4  #include <stdlib.h>
 5
 6  int* buggy_fn();
 7
 8  int main(int argc, char *argv[]){
 9
10   // atoi() does not check for errors.
11   // Watch what happens when we give an error of omission.
12   int array_size = atoi(argv[1]);
13
14   int *array;
15   array = buggy_fn(array_size);
16
17   int index;
18
19   for(index=0; index<array_size; index++){
20   printf("array[index] = %d\n", array[index]);
21   }
22
23   return 0;
24  }
```

FIGURE 5.1
Buggy program for debugging.

```
 1 // buggy_fn.c
 2
 3 #include <stdio.h>
 4
 5 // This function returns a pointer to an array
 6 // that was allocated within the function.
 7 // Guess what happens to the array
 8 // when the function returns!
 9
10 int *buggy_fn(int array_size){
11
12   int index, array[array_size];
13   for (index=0; index<array_size; index++){
14   array[index] = index;
15   }
16
17   int *retval;
18   retval = array;
19
20   for (index=0; index<array_size; index++){
21     printf("retval[index] = %d\n", retval[index]);
22   }
23
24   return retval;
25 }
26
```

FIGURE 5.2
Buggy function.

```
1       #!/usr/bin/make
2
3       DEBUG:= -O0 -g
4
5       default:
6           @echo "making with DEBUG=$(DEBUG)"
7              make buggy
8
9       buggy: buggy.c buggy_fn.o
10          gcc $(DEBUG) buggy.c buggy_fn.o -o buggy
11
12      buggy_fn.o: buggy_fn.c
13          gcc $(DEBUG) -c buggy_fn.c
14
15      clean:
16          rm -f buggy_fn.o buggy
```

FIGURE 5.3
Makefile for building the buggy program.

text into an integer type but has fallen out in favor of the `strtol()` function, which does check for errors. We use it to illustrate perhaps the most common type of bug: Segmentation fault resulting from an attempt to de-reference a null pointer.

The second is another common bug. Here, `buggy _ fn()` returns a pointer to an array that was allocated within the function itself. This is a bug because when the function returns, that memory that was allocated for its variables no longer belongs to that function. Trying to access memory that is no longer allocated for you is like going into your old apartment after you have already moved out. Unexpected results can occur! You might find it just as you left it (memory unchanged and giving you result you expect), you might find someone else has moved in (memory changed in some unpredictable way), or you might encounter a more unpleasant circumstance (segmentation fault, for trying to access memory that no longer belongs to you). The program may compile and run and behave differently at different times, or when using different compiler options, making it difficult to pin down. We will walk through it with the debugger to show when things *might* get out of hand (Figures 5.2 and 5.3).

5.3 Instrumenting the Code

Before running the debugger, we must prepare the code for debugging. Fortunately, this is handled by the compiler and does not require any re-working of the code. To generate the debug version of our code, we supply the compiler with two options:

```
gcc -g -O0 buggy.c buggy_fn.o -o buggy
```

-g

Passing this option to the compiler tells it to instrument the code for debugging. Without properly instrumenting the code, the debugger has no way to connect with the code.

Instrumenting the code is just a fancy way of saying that you are asking the compiler to include human-readable stuff such as names, line numbers, and the code itself, so that you can refer back to the original program. When you compile code for production, this stuff is left out so that it is more efficient.

-O0

This option tells the compiler not to do any optimizations (literally, it tells it to use optimization level 0). Modern C compilers are capable of so much more than merely translating C code into machine code. Higher optimization levels will do things such as unroll loops (write each iteration explicitly, faster than jumping back and incrementing a counter), rearrange the order of instructions (sometimes code can be made to run faster and deliver equivalent results), and delete instructions (e.g., if their result is never used) and all sorts of other black magic that has been determined not to affect results, but that will distinctly not look as the same code that was written. So to keep things readable during debugging, make sure to include this flag and tell the compiler not to optimize.

5.4 Running the Program without the Debugger

So you have compiled and instrumented your code and you were careful not to include any optimizations. You now have a slow, bloated debugger-friendly code that you execute, so

```
$ ./buggy
Segmentation fault
```

Ouch! We just got started and already a segmentation fault? Remember that in describing the code, we stated that it would take one command line argument, which would be the length of the array to be generated. Since we did not specify a value, `atoi()` received "null" and the program crashed and died. What is to be done? Turns out that `gdb` can do a postmortem on a dead program, and this will be our point of entry, but first, `gdb` needs one more thing: a core dump.

5.5 Core Dump

When a program terminates abnormally, the OS will attempt to "dump core" if permitted. A core dump is a snapshot of what the program looked like in memory when it was executing. When a program dies unexpectedly, the core contains information where and on the values of variables at that point. This is pretty handy information for an autopsy of the dead process.

To examine a core dump, we first need to get a core file. Core files are large, and many system administrators will set up their systems so as not to allow core dumps by default, lest they receive calls from users asking why they ran out of disk space. To specifically allow core dump on your system, try:

```
$ ulimit -c unlimited
```

And then run the program. It will generate the core file when it hits the segmentation fault. If this does not work, contact your system administrator. Names may vary and often include the process ID as part of the name. For this example, the core file is just called core. We open the core file with the debugger, by specifying the name of the program that generated it:

```
$ gdb buggy core

...
Reading symbols from buggy...done.
[New LWP 6619]
Core was generated by './buggy'.
Program terminated with signal SIGSEGV, Segmentation fault.
#0  0xb75cfa62 in __GI_____strtol_l_internal (nptr=nptr@entry=0x0,
endptr=endptr@entry=0x0, base=base@entry=10, group=group@entry=0,
      loc=0xb77468a0 <_nl_global_locale>) at strtol_l.c:298
298     strtol_l.c: No such file or directory.
```

5.6 Backtracing

The debugger tells us that the program terminated at line 298 of `strtol_l.c`, which we do not recognize because it is not our code. In fact, the debugger tells us "no such file or directory" because it has no information for that line. This is because in this case, the fault occurred at some code in the standard library. We called a function from the standard library that crashed the program! Incidentally, this is why using `atoi()` is not recommended. But we just want to know where in *our* code this is happening. To do this, we use backtrace:

```
(gdb) backtrace
#0  0xb75cfa62 in __GI_____strtol_l_internal (nptr=nptr@entry=0x0,
endptr=endptr@entry=0x0, base=base@entry=10, group=group@entry=0,
   loc=0xb77468a0 <_nl_global_locale>) at strtol_l.c:298
#1  0xb75cf827 in __GI_strtol (nptr=nptr@entry=0x0, endptr=endptr@
entry=0x0, base=base@entry=10) at strtol.c:108
#2  0xb75cca9f in atoi (nptr=0x0) at atoi.c:27
#3  0x08048466 in main (argc=1, argv=0xbfef3ee4) at buggy.c:12
```

This lists the stack right back up to the place in our program where it was called. We see line 12 in `buggy.c` (our code) made this call. Do not remember what line `buggy.c:12` is about? `gdb` will list code contextually for you:

```
(gdb) list buggy.c:12
7
8       int main(int argc, char *argv[]){
9
10       // atoi() does not check for errors.
11       // Watch what happens when we give an error of omission.
12       int array_size = atoi(argv[1]);
13
14       int *array;
15       array = buggy_fn(array_size);
16
```

If you do *nothing else* with gdb, at least know that you can pinpoint the exact location where the program failed. Bisecting code with endless print statements also works, but again, you have the appropriate tools, so use them.

5.7 Invoking the Code with the Debugger

So, now you know the first thing about using the debugger. Let's do something fancy now. Let's run the program again, invoking it with the debugger. We start the debugger by specifying the executable "buggy":

```
$gdb buggy
(gdb)
```

5.8 Setting (and Deleting) Breakpoints

A breakpoint is a place where we want the code to stop or "take a break" from execution. Notice that we still have not specify the command line argument, and we know that it is going to crash when it makes that call to `atoi()`, so we set a breakpoint right before it, at line `buggy.c:12`:

```
(gdb) break buggy.c:12
Breakpoint 1 at 0x8048456: file buggy.c, line 12.
```

Then we use the start command to run the program up to this breakpoint:

```
(gdb) start
Temporary breakpoint 2 at 0x8048456: file buggy.c, line 12.
Starting program: /home/willmore/gdb-chapter/buggy

Breakpoint 1, main (argc=1, argv=0xbffff014) at buggy.c:12
12          int array_size = atoi(argv[1]);
```

Notice that the debugger has already added a temporary breakpoint in the same location. This is because this is the first line of code to be executed. The debugger will automatically set a break at the first line to execute. The program has stopped executing at the breakpoint and now waits for our instructions. One of the great features of gdb is that you can examine and even modify the values of variables! In this case, we look at the value of argv[1], the command line argument that we neglected to pass, using the gdb print (abbreviated p) command:

```
(gdb) p argv[1]
$1 = 0x0
```

5.9 Modifying Variables

Behold, argv[1] contains a null pointer, since we passed no value! Since the program is stopped, we can modify this value here. Let's change it to "4" (not 4.). We are using a string value here because argv[] stores the values received at the command line, which are strings.

```
(gdb) set variable argv[1] = "4"
```

and in case we do not believe it worked, examine it again:

```
(gdb) p argv[1]
$2 = 0x804b008 "4"
```

Great! The debugger lists each line of code that is about to be executed. We set an additional breakpoint at the first point after the atoi()call and tell gdb to continue executing:

```
(gdb) break 14
Breakpoint 3 at 0x804846a: file buggy.c, line 14.
(gdb) continue
Continuing.

Breakpoint 3, main (argc=1, argv=0xbffff014) at buggy.c:15
15          array = buggy_fn(array_size);
```

Notice that we did not specify which file contains the line. If no file is specified, the current file is assumed. Notice also that line 14 only lists a declaration, no instructions. The debugger actually goes past this line and stops at the next instruction to be executed, which is on line 15. We can now look at the value of array _ size, received from atoi():

```
(gdb) p array_size
$1 = 4
```

We see the value 4 (not "4") has been recovered. We have successfully modified a value in a running program! Powerful stuff! To allow the program to finish executing, we again type continue:

```
(gdb) continue
Continuing.
retval[index] = 0
retval[index] = 1
retval[index] = 2
```

```
retval[index] = 3
array[index] = 0
array[index] = 805306369
array[index] = 2
array[index] = 3
[Inferior 1 (process 7509) exited normally]
```

5.10 Modifying Breakpoints

And we see that the program exits normally. Never mind for now the garbage result that it generates, we will get to that. The good news is that while it is possible, you do not need to specify command line parameters this backward way. We clear the breakpoints and run the program again, specifying the command line parameter with the *run* command:

```
(gdb)   info breakpoints
Num     Type           Disp Enb Address    What
1       breakpoint     keep y   0x08048456 in main at buggy.c:12
        breakpoint already hit 1 time
3       breakpoint     keep y   0x0804846a in main at buggy.c:14
        breakpoint already hit 1 time
(gdb)   delete 1 3
(gdb)   info breakpoints
No breakpoints or watchpoints.
(gdb)   run 4
Starting program:/home/willmore/gdb-chapter/buggy 4
retval[index] = 0
retval[index] = 1
retval[index] = 2
retval[index] = 3
array[index] = 0
array[index] = 805306369
array[index] = 2
array[index] = 3
[Inferior 1 (process 7521) exited normally]
```

The program runs the same. We are left with the information that `atoi()` is deficient (it generates a segmentation fault rather than recording and reporting an error condition) and should be replaced. Also, we see that something is amiss with the rest of the program.

5.11 Listing and Stepping through Code

Let's set a new breakpoint and then start the code again from the beginning:

```
(gdb) break 15
Breakpoint 5 at 0x804846a: file buggy.c, line 15.
(gdb) run 4
Starting program: /home/willmore/gdb-chapter/buggy 4
```

```
Breakpoint 5, main (argc=2, argv=0xbffff014) at buggy.c:15
15          array = buggy_fn(array_size);
```

Lovely. The code has run up to the (break)point of interest. Let's start stepping through the code one line at a time. We really want to focus in here:

```
(gdb) step
buggy_fn (array_size=4) at buggy_fn.c:12
12          int index, array[array_size];
```

Note that if you ever lose track of where you are, you can just ask where:

```
(gdb) where
#0  buggy_fn (array_size=4) at buggy_fn.c:12
#1  0x08048476 in main (argc=2, argv=0xbffff014) at buggy.c:15
```

And you can get the code listing (few lines before and after) with list:

```
(gdb) list
7      //Guess what happens to the array
8      //when the function returns!
9
10     int *buggy_fn(int array_size){
11
12         int index, array[array_size];
13         for (index=0; index<array_size; index++){
14         array[index] = index;
15         }
16
```

We are on line 12 in buggy.c, which looks like just a declaration but also contains a stack allocation request for "array." Let's continue stepping through the code. Note that if you just hit enter, without entering a new command, gdb will keep executing the last command, in this case "step":

```
(gdb) step
13          for (index=0; index<array_size; index++){
(gdb)
14           array[index] = index;
(gdb)
13          for (index=0; index<array_size; index++){
(gdb)
14           array[index] = index;
(gdb)
13          for (index=0; index<array_size; index++){
(gdb)
14           array[index] = index;
(gdb)
13          for (index=0; index<array_size; index++){
(gdb)
14           array[index] = index;
(gdb)
13          for (index=0; index<array_size; index++){
```

```
(gdb)
18          retval = array;
(gdb) p index
$2 = 4
```

This carries us through four iterations of the loop. Note that the loop exits because the test condition (`index<array _ size`) is no longer true. You can examine any value at any point in the loop. The next loop attempts to check the work (using print statements!) before returning. We can step through the code until our heart is content. We will see that the program is behaving as we would expect it to behave while in the function.

5.12 Running from Breakpoint to Breakpoint

Let's take a step back and examine the main program. Since the code appears to be behaving normally within the function, let's set a breakpoint in the main program immediately following the function. Better yet, let's exit the debugger and start clean. And this time, we will execute the program with *run* instead of *start*, which doesn't stop at the first instruction and will instead run to the first breakpoint:

```
$ gdb buggy
(gdb) break 18
Breakpoint 1 at 0x804847a: file buggy.c, line 18.
(gdb) break 20
Breakpoint 2 at 0x8048484: file buggy.c, line 20.
(gdb) run 4
Starting program: /home/willmore/gdb-chapter/buggy

retval[index] = 0
retval[index] = 1
retval[index] = 2
retval[index] = 3

Breakpoint 1, main (argc=2, argv=0xbffff014) at buggy.c:19
19          for(index=0; index<array_size; index++){
```

We have reached the first breakpoint, after the function return. Let's examine the values of our array:

```
(gdb) p array[0]
$1 = 0
(gdb) p array[1]
$2 = 1
(gdb) p array[2]
$3 = 2
(gdb) p array[3]
$4 = 3
```

Things appear as we hoped they would. Let's continue running to the next breakpoint, which lies inside the loop that prints the returned values:

```
(gdb) continue
Continuing.

Breakpoint 2, main (argc=2, argv=0xbffff014) at buggy.c:20
20          printf("array[%d] = %d\n", index, array[index]);
(gdb) p index
$5 = 0
(gdb) p array[0]
$6 = 0
(gdb) p array[1]
$7 = 1
(gdb) p array[2]
$8 = 2
(gdb) p array[3]
$9 = 3
```

First pass through the loop and everything still appears as we thought. Continue the loop through the next iteration, and examine values again at the next breakpoint:

```
(gdb) continue
Continuing.
array[0] = 0

Breakpoint 2, main (argc=2, argv=0xbffff014) at buggy.c:20
20          printf("array[%d] = %d\n", index, array[index]);
(gdb) p array[0]
$10 = 0
(gdb) p array[1]
$11 = 805306369
(gdb) p array[2]
$12 = 2
(gdb) p array[3]
$13 = 3
```

Aha! Very interesting! The values of `array[]` have changed, but there is no line of code that has been executed that specifically asks for these values to be modified. We have drilled down into it with the debugger and can say with certainty that this is true. This is a difficult class of bug to track down, as it reflects modification of memory by something outside of the instruction stream. By eliminating the possibility of the issue arising from the instruction stream, we can make an informed decision that an asynchronous process (e.g., another thread in a parallel code or a process from the operating system itself) is the source of the unexpected result. I would like to say that most bugs are simpler than this, but as you begin programming at more and more sophisticated levels, the bugs get subtler. The error here is in attempting to reference a stack variable outside of the scope it was declared. A potential solution may be to allocate the desired array from the heap instead.

5.13 Conditional Breakpoints

There are many layers of sophistication attainable in gdb, and I will leave you with just one more. It is possible and often desirable to set a breakpoint based on a certain logical condition; for example, we want to stop our program when the array value is not equal to its index. gdb allows for this explicitly:

```
(gdb) break 20 if (array[index] != index)
Breakpoint 1 at 0x8048484: file buggy.c, line 20.
(gdb) run 4
Starting program: /home/willmore/gdb-chapter/buggy 4
retval[index] = 0
retval[index] = 1
retval[index] = 2
retval[index] = 3
array[0] = 0

Breakpoint 1, main (argc=2, argv=0xbffff054) at buggy.c:20
20          printf("array[%d] = %d\n", index, array[index]);
(gdb) p index
$1 = 1
(gdb) p array[index]
$2 = 805306369
```

Lo and behold, the code stops where the discrepancy arises! Applying the condition in this case requires knowing the answer to our question beforehand; nevertheless, it can be a powerful tool, and these are only a few of the options. gdb is part of the GNU toolchain and additional information on gdb can be found at http://www.gnu.org/software/gdb/documentation/.

6

Makefiles, Libraries, and Linking

Frank T. Willmore

CONTENTS

6.1 Makefiles

Software is inherently complex, and the management of its development requires special tools. This task is often called build management, or build engineering, and is one of the most powerful tools available for it is the standard utility 'make'. Through the use of `makefiles`, `make` provides a way of efficiently codifying and organizing the instructions to be executed when building a program.

You may have, at some point, received a short Bash script or one-liner that you used to build a program, almost like a spell or incantation that would cause the program to magically compile and link and maybe even execute. It is time to deepen our understanding. Take this simple program example, written in C:

```
1    /* c_program.c */
2
3    #include <stdio.h>
4    #include <stdlib.h>
5
6    #include "c_function.h"
7
8    int main(int argc, char **argv){
9
10       int val = strtol (argv[1], NULL, 10);
11       printf("Running C program with val = %d\n", val);
12
13       c_function(&val);
14       printf("c_function returns with val = %d\n", val);
15
16       printf("C program complete.\n\n");
17       return 0;
18   }
```

The C-language header file containing the function prototype:

```
1    /* c_function.h */
2
3    void c_function(int *);
```

The C-language file containing the function implementation:

```
1    /* c_function.c */
2
3    #include "c_function.h"
4
5    /* multiply the input value by 2 */
6    void c_function(int *val){
7       *val = *val * 2;
8    }
```

And a script, which will compile and run the program:

```
1    #!/usr/bin/env bash
2    gcc -c c_function.c
3    gcc -c c_program.c
4    gcc c_function.o simple_program.o -o c_program
5    ./c_program 1
```

Let's run this script line by line and look at what's happening. In the most general case, you will have a program (e.g., `c _ program.c`), which will call one or more components (`c _ function` in this case) and will be built in many steps. The first step, as shown, is to compile the function. The compiler flag '-c' tells the compiler to compile, but not to link:

```
$ ls
c_function.c c_function.h c_program.c
$ gcc -c c_function.c
$ ls
c_function.c c_function.h c_function.o c_program.c
```

The output when using this option is a file with the same prefix but the '.o' suffix. This program has two components: The `c _ program`, which contains the main() function, and the `c _ function` it calls. Each is compiled individually:

```
$ gcc -c c_program.c
$ ls
c_function.c c_function.h c_function.o c_program.c c_program.o
```

Line 4 of the script (does not contain '-c') takes the two '.o' files and links them together to form the output ('-o' option) `c _ program`:

```
$ gcc c_function.o simple_program.o -o c_program
$ ls
c_function.c c_function.h c_function.o c_program c_program.c c_program.o
```

The final line of the script runs `c _ program` with the command line argument '1':

```
$ ./c_program 1
Running C program with val = 1
c_function returns with val = 2
C program complete.
```

This simple code could have been written in a single file and compiled with a single command. Most codes are not so simple. Most research codes are the product of many collaborators, contain many components, and are written with testing and extensibility in mind, and we break out the individual steps here to illustrate the level of detail that should go into designing a good build system. The design, implementation, and maintenance of a sound mechanism to manage development, dependencies, and compilation, sometimes across multiple platforms, are called build engineering. A good makefile can be an important part of the build engineering design for a given code.

Let's see what a makefile might look like for the same project. We will take the build commands from the script above, show what they would look like as a makefile, then add some details that make it work better than the script. To get started, we introduce the idea of *targets*. A target is something that the makefile understands how to make, and every makefile should have at least one target. For `c_program`, one of the targets would be the '.o' or object file, and the way we would specify that target would be

```
c_program.o: c_program.c
	gcc -c c_program.c
```

`c_program.o` is the target. It appears to the left of the colon. The things on which the target depends are called *dependencies* and appear as a list to the right of the colon. The second line is the *rule*. The rule gets executed when you ask make to update a target, but only if the dependency has been modified more recently than the target.

Let's say that you have a makefile with just the above two lines. If you execute a make command to update `c_program.o`:

```
$ make c_program.o
```

Here's what happens: make will look to see if the dependency (`c_program.c`) has been updated more recently than the target (`c_program.o`). If it has, say because we edited `c_program.c`, then make will execute the rule, causing `c_program.o` to be updated, in this case by invoking `gcc` to recompile it. By including targets as dependencies for other targets, cascading dependencies can be specified in the makefile, so that every bit that needs updating gets updated when its dependency gets updated, but without having to rebuild every single piece of a codebase every time we wish to update its targets. Some large codes can take hours and even days to compile, so save yourself and everyone who depends on you, time and write the right dependencies.

Let's see what a good makefile might look like, by implementing all the rules of our script, plus a few extras:

```
#!/usr/bin/env make
# Beginning a line with a '#' makes it a comment.
# Comments are great for leaving notes for yourself
# and anyone else who might read your Makefile

# If you don't tell make what target to build (you just type 'make')
# then it will attempt to update the first target in the file.
# I specify a default target, just in case I don't want it to run,
# say, what would be the first target 'clean' which will delete stuff.
# Deleting stuff is scary.
```

```
# Specify a compiler using a macro. $(CC) will be substituted with gcc
# wherever $(CC) is found in the Makefile.
CC:=gcc

default:
	@echo "No target specified."

# A lot of people will specify a target called 'clean' which will
# clean up intermediary stuff, e.g., object files. Notice that clean
# does not depend on anything. This means, if you type 'make clean'
# then the rule will always get executed. It is sort of like an alias,
# but specific to the Makefile and directory you are in.

clean:
	rm -f *.o c_program

# A Makefile can be used to run a program. It could be used to run the
# program you are building (like this one) or it could be used to run
# another program. It could be used to run *any* program. It could be
# used to send an email or text, to call a phone, or to run a program
# that will transfer all of your assets to a bank in Singapore. I
# worry sometimes that someone in New York might have a target in a
# Makefile called 'global_financial_chaos'. I worry about dumb stuff.

run: c_program
	./c_program

# This is the rule to build the c_function.o object. Straight from the
# script.

c_function.o: c_function.c c_function.h
	$(CC) -c c_function.c

# This is the rule to compile, but not link c_program.

c_program.o: c_program.c
	$(CC) -c c_program.c

# This is the rule to link the object files in order to produce
# c_program. Again, just as in the script.

c_program: c_program.o c_function.o
	$(CC) c_program.o c_function.o -o c_program

# .PHONY is a way to say "I want to use these names for targets.
# There are no files associated with them, I just want to be able to
# say 'make run' and 'make clean' and have it do stuff." If a file
# called 'run' or 'clean' exists, make will just ignore them, because
# you let it know that you mean something else.

.PHONY: run clean
```

That's it. We will add a few more bits later, but at its simplest, a makefile is a list of dependencies and rules for updating targets, with comments. I included a lot of comments, some might say too many. Most codes have too few. Err on the side of including too many comments. Your collaborators will benefit, even if they complain about it. Actually, I have never heard anyone complain about too many comments.

You should be able to write a simple makefile now. A very important detail to remember is that the whitespace before the update rule (the indented lines following a dependency) is a tab (ASCII character 0x09). It is *not* five spaces, eight spaces, and so on, and if you are copying and pasting (e.g., from this book) to create a makefile, be aware that it may not be copied correctly. Always use a tab!

6.2 Libraries

Now that we are experts in using a makefile to build a simple program with a single function call, let's introduce function libraries, which you have most certainly been using, but which you may not yet have written. A canonical example of a function library is the standard math library, which contains implementations of standard mathematical functions (square root, exponential, etc.) and whose interface is defined in the commonly #included math.h header file.

Some characteristics of a function library are the following:

- Its interface is defined in a header file.
- It exists as either an .a (static) or .so (dynamic/shared) binary file in a library directory.
- It is built from one or more .o binary files.
- It is named libxxx.so and lives in a .../lib directory.
- It is referenced by a compile command with the -l and -L options.

As an example, we will extend the above code to include a second C function c_function_d, which does the same thing as c_function, but for a double, and we will put both of these functions into both a static (.a) and a dynamic (.so) library.

First, the function definition (prototype):

```
1   /* c_function_d.h */
2
3   void c_function_d(double *);
```

The C-language file containing the function implementation:

```
1   /* c_function_d.c */
2
3   #include "c_function_d.h"
4
5   /* multiply the input value by 2.0 */
6   void c_function_d(double *d_val){
7       *d_val = *d_val * 2.0;
8   }
```

We add a target to the makefile:

```
c_function_d.o: c_function_d.c c_function_d.h
	$(CC) -c c_function_d.c
```

And a line to build the static (.a) library:

```
libmy_c_functions.a: c_function.o c_function_d.o
	ar -cvq libmy_c_funcions.a c_function.o c_function_d.o
```

To use the library, we add the following call to `c_program.c`:

```
#include "c_function_d.h"
...
double d_val = (double)val;
c_function_d(&d_val);
printf("c_function_d returns with d_val = %lf\n", d_val);
```

And we no longer need the `.o` files, since their functionality is contained in the library. The compile command for `c_program` becomes

```
$(CC) c_program.c -lmy_c_functions -L.-o c_program
```

We are no longer including the step of compiling `c_program.c` to produce the intermediary `c_program.o` object file. The only object files used are the ones in the library, referenced with the `-lmy_c_functions` option. The compiler expects specifying -lxxx at the command line to mean 'look for a library archive named libxxx.a or libxxx.so in one of the places it knows to look for libraries.' The -L option tells it to look in the directory included with the option. In this case, -L. tells the compiler to look in '.' that is, the current directory. Note that if your code requires -L{some directory} at compile time in order to find a given shared library, then the environment variable LD_LIBRARY_PATH will need to include {some directory} at runtime.

Since the library we built was a static library, it means that the entire contents of the .a file will be included in the executable and loaded into memory when the program is loaded into memory, in preparation for it to be run. Whether to use a static or dynamic executable is a design decision. The biggest difference between using static and dynamic libraries is that *statically linked* code is incorporated directly into the executable. This means that statically linked executables will be larger than *dynamically linked* ones. It also means that the library calls will be loaded and resident in memory, so that when calls are made to them, there is no waiting for the library to be loaded. The downsides are that larger executables take longer to load, and loading multiple programs (or instances of programs) that use them means that each will have its own copy loaded in memory, consuming more system resources. Dynamic libraries are loaded as needed (i.e., when they are actually called by the code) and can even be loaded explicitly using `dlopen()`. They are only slightly more complicated to use and are generally the best choice. One other practical concern that can arise is when a code consists of multiple components that depend on different versions of the same dynamic library but can sometimes be built by linking static versions of the needed libraries for different components. This can result from either poor build management or from the integration of code bases from different projects.

Building a dynamic library is very similar to building a static library but requires a few additional options in the compile commands, namely, that -fPIC is specified when compiling each component (stands for position independent code):

```
c_function.o: c_function.c c_function.h
	$(CC) -fPIC -c c_function.c

c_function_d.o: c_function_d.c c_function_d.h
	$(CC) -fPIC -c c_function_d.c
```

and the library itself is created from the .o files by the compiler using the -shared option:

```
libmy_c_functions.so: c_function.o c_function_d.o
	$(CC) -shared -o libmy_c_functions.so c_function.o c_function_d.o
```

Building the c _ program itself uses the same compile command:

```
$(CC) c_program.c -L. -lmy_c_functions -o c_program
```

The GNU C compiler will choose the .so by default if both the .so and .a versions exist. An important tool for determining which shared libraries are used by a program is ldd (list dynamic dependencies), which will display the names of the dynamic libraries on which it depends. These typically include standard system libraries, the standard C library (libc), as well as user-defined libraries:

```
$ ldd c_program
    linux-gate.so.1 =>  (0xb76ec000)
    libmy_c_functions.so => ./libmy_c_functions.so (0xb76e6000)
    libc.so.6 => /lib/i386-linux-gnu/libc.so.6 (0xb7523000)
    /lib/ld-linux.so.2 (0xb76ed000)
```

Notice the dependence on our libmy _ c _ functions.so for the executable built with dynamic linking. By contrast, the statically linked version does not:

```
$ make clean
$ make -f Makefile.static c_program
$ ldd c_program
    linux-gate.so.1 =>  (0xb7783000)
    libc.so.6 => /lib/i386-linux-gnu/libc.so.6 (0xb75bd000)
    /lib/ld-linux.so.2 (0xb7784000)
```

6.3 Linking

To summarize linking, consider the nm command, which allows the user to view symbols (functions, global variables, etc.) in an object file. For our c _ function.o:

```
$ nm c_function.o
00000000 T c_function
```

nm tells us that the object file contains one symbol, `c _ function`, and the `T` tells us that it is defined in the text/code section. Running `nm` on executables and shared libraries will deliver a lot more information about global variables and system info that gets rolled in when linking. For example, filtering the contents for `c _ program` to look only for symbols of type 'T', we see that the familiar function 'main' is defined, as well as our functions 'c_function' and 'c_function_d':

```
$ nm c_program | grep " T "
08048523 T c_function
0804853a T c_function_d
080485c4 T _fini
080482f4 T _init
080485c0 T __libc_csu_fini
08048550 T __libc_csu_init
0804847d T main
08048380 T _start
080483b0 T __x86.get_pc_thunk.bx
```

This will be the output for the executable that was built from either the static (`.a`) or directly from the object (`.o`) files. Notice that when the executable is built using the shared library, that nm lists our functions as undefined "U":

```
$ nm c_program |grep "c_fun"
         U c_function
         U c_function_d
```

Again, we see that dynamically linked functions are defined outside of the executable.

So, the ldd utility shows what libraries an executable (or another library) depends upon to run, and nm gives information on specific functions that are either implemented in the file or are expected to be implemented elsewhere.

You should now be able to create basic makefiles to manage build for your projects and be able to understand and possibly modify other projects that use make to manage build. You now also have a basic understanding of how libraries are built and codes are linked, how to examine binaries using nm, and how to check dynamic library dependencies using ldd.

GNU make is part of the GNU toolchain and additional information can be found at https://www.gnu.org/software/make/.

7

Linking and Interoperability

Frank T. Willmore

CONTENTS

Great work leverages other great work. However, that great work is not always in the same language. In coding, it is no different. There are powerful numerical libraries and simulators written in Fortran, powerful analytical tools written in Python, and great molecular mechanics programs written in C and C++. Some languages sometimes work better for some people for some things. Fools debate the merits of one paradigm over another and fail by bemoaning what should or should not be done. The wise understand the nuances and differences and succeed by getting them to work together. Here, we teach you to be wise.

Building on what we have learned about linking from our previous chapter and C-language example, we first show how to directly link codes compiled from different languages, namely, C++ and Fortran. After this, we introduce simple Boost::python calls, which allow C++ code (and, therefore, C and Fortran code, by extension) to be called from Python.

7.1 Fortran Calling Fortran

We have seen C calling a C function. In the Fortran paradigm, the function call looks like this:

```
! fortran_program.f90

program fortran_program
  implicit none
  integer, external :: fortran_function
integer::val
integer::return_val
val = 1

  print *, "Running Fortran program with val = ", val
return_val = fortran_function(val)
```

```
  print *, "fortran_function returns val = ", val
  print *, "Fortran program complete."
  print *, ""

end program fortran_program
```

And the Fortran function looks like this:

```
! fortran_function.f90

integer function fortran_function(val)
  implicit none
  integer val
  val = val * 3
  fortran_function = 0
end function fortran_function
```

Just as was possible with C, each of these components can be compiled separately, then linked:

```
$ gfortran fortran_function.f90 -c
$ gfortran fortran_program.f90 -c
$ gfortranfortran_function.ofortran_program.o -o fortran_program
```

It might make sense, then, that we might be able to mix and match components, and this is in fact the case. Examining the contents of the fortran_function.o using nm, we see that the fortran compiler appends an underscore (_) to the end of the function name:

```
$ nm fortran_function.o
00000000 T fortran_function_
```

Could making these two languages work together be as simple as adding an underscore to the end of a C function's name? The short answer is yes; it can be just that simple.

7.2 Fortran Calling C

Let's begin with the C function we wrote in the previous chapter, the one that multiplies by 2 and returns. Something that needs to be noted about that function is that it is written to pass by reference, not by value. The reason for this is, in Fortran, *all* functions pass by reference, so for any function call to be compatible with Fortran, it must also pass by reference. This does not mean every function in the code needs to pass by reference, just the ones that interact with Fortran. So, to make our c_function callable from Fortran, we add a wrapper function called c_function_().

```
/* c_function.h */

// original function
intc_function(int *);

// fortran wrapper
intc_function_(int *);
```

The wrapper function name is the same as the original but appends an underscore ( ) to the end of the name. The reason for this is, when a Fortran function is compiled, the compiler appends an underscore to the end of the name. Thus, the first step in making c_function callable from Fortran is to make it look like a Fortran function. Note that if you were writing a *new* C function to be called from Fortran, you would just name it with the trailing underscore. We are showing how to wrap an *existing* C function here.

We keep the exact same function definition as before for the c_function, but we need to define the Fortran wrapper in c_function.c:

```
#include "c_function.h"

intc_function(int *val){
  *val = *val * 2;
  return 0;
}

// Fortran wrapper
intc_function_(int *val){
  return c_function(val);
}
```

It takes the argument and calls the original c_function. Simple! To include this function to our Fortran program, we add the following to the declarations at the beginning of the program:

```
integer, external :: c_function
```

This takes the place of the header file with function prototype, which Fortran does not use. To invoke the function, we add the following call, and then examine the output with a print statement:

```
return_val = c_function(val)
print *, "c_function returns val = ", val
```

Take the c_function.o object file from the C compiler and the fortran_program.o object file from the Fortran compiler and link them together. Brilliant! But with which compiler do we do the linking? Remember that the final linking to make an executable requires certain system stuff to be handled, so whichever compiler was used to write the *main program* should here be used to do the final linking. In this case, it is the Fortran compiler:

```
$ gfortranc_function.ofortran_program.o -o fortran_program
```

Because the function arguments to both the C and Fortran functions are passed by reference, the effect of calling c_function(), which multiplies by 2, and fortran_function(), which multiplies by 3, is cumulative. The final value of val is 6× its original value:

```
$ ./fortran_program
Running Fortran program with val = 1
c_function returns val = 2
fortran_function returns val = 6
Fortran program complete.
```

7.3 C Calling Fortran

Add a trailing underscore to a C function name and Fortran calling C works like magic! What about the reverse trick? It should be easy enough to find out. Because we already have the functions written, we should just be able to add the call to fortran_function() to our C program. But remember that funny thing the Fortran compiler does, adding that trailing underscore to the function name. Could calling the Fortran function be as easy as adding the underscore and passing by reference when making the call?

```
/* C code */
fortran_function_(&val);
printf("fortran_function returns with val = %d\n", val);
```

Compile and link, this time using the C compiler:

```
gcc c_program.c -c
gcc c_program.ofortran_function.o -o c_program
```

Yes, it is this easy! And again, we see the result we expect. Language interoperability works and is relatively straightforward. Add the trailing underscore and remember to pass by reference.

7.4 And Beyond. C++ Really!?

People with strong opinions about the suitability of one language versus another may bring up the importance of objects and modularity and many other issues. For those of us who just want to get our code to work, just know that C++ can be linked as well. We are going to build this little program in C++, C, and Fortran and we are going to write the functions each way, too, including both as C-style functions in C++ as well as static class methods. We will not get into passing objects around. I am promising you a road, not a freeway. You are going to have to write some code in a language you do not like at some point, even if just to build a bridge to your own code, and I am giving you the tools to do so. Here is what you need to know about interoperating with C++.

It bears mentioning at this point that although C and C++ are often spoken of together (it is common to see C/C++ written as a single skill on a resume), there are some important differences. While it is true that code written in C can generally be read and compiled by a C++ compiler and that standards committees for both languages work very hard to make sure this is the case, know that there are profound differences in how code is written and that a C++ compiler is not simply a fancy C compiler. Know which language you are writing, and know that the compilers will do some very different things under the hood.

Consider name mangling, which is specific to the C++ compiler. That funny little thing that the Fortran compiler does, putting the trailing underscore, is nothing compared to what the C++ compiler does to a function. In a nutshell, name mangling is what the C++ compiler does because C++ allows overloading; that is, more than one function can have the same name, because one might want to write a function that does something similar to different types of arguments, for example, an int version, a float version, and a complex version. The bottom line is that the C++ compiler makes the function name come out funny, and you have to work around it. To see an example, take the following C++ header:

```
// cpp_function.hh
intcpp_function(int *);
```

and function:

```
// cpp_function.cpp
#include "cpp_function.hh"

intcpp_function(int *val){
  *val = *val * 5;
  return 0;
}
```

After compiling to generate cpp_function.o, we use nm to see:

```
$ nm cpp_function.o
00000000 T _Z12cpp_functionPi
```

This tells us that in order to make the function name unique (for different types of argument list) C++ compensates by mangling the function name in a unique way. The function name is still in there, but there is extra stuff that makes it unique. But how do we know what to call? That is more complicated than just adding an underscore. Fortunately, C++ provides a solution in the form of externing.

7.5 External Linkage/Externing

Name mangling can be avoided by *externing*. This is not to be confused with *extern*, a keyword applied as a *variable* modifier, which indicates that the variable is to be allocated outside of the current source file. Declaring a function or block of code as *"extern C"* in a C++ file indicates that a C-style linkage should be generated for the given function(s). When C or C++ code is complied with the C++ compiler, the macro "__cplusplus" is defined, allowing the preprocessor to handle certain bits of code differently. In this example, we have used *"extern C"* to ensure that our C++ functions cpp_function() and cpp_function_() will not have their names mangled by compilation, so that they can be more easily called from C, or Fortran:

```
// cpp_function.hh

#ifdef __cplusplus
extern "C"
{
#endif
  intcpp_function(int *);
  intcpp_function_(int *);
#ifdef __cplusplus
}
#endif
```

Implementation is as follows:

```
// cpp_function.cpp

#include "cpp_function.hh"

intcpp_function(int *val){
  *val = *val * 5;
  return 0;
}

// fortran wrapper
intcpp_function_(int *val){
  return cpp_function(val);
}
```

Thus, we have exposed a C++ function to C and Fortran. Now let's go further and generate a function with an unmangled function name so that your C/Fortran code can call functions with mangled names, for example, the static methods on a C++ class. See this work in the following prototype:

```
// cpp_class_method_wrapper.hh

#ifdef __cplusplus
extern "C"
{
  intcpp_class_method_wrapper(int *);   // for C
  intcpp_class_method_wrapper_(int *);  // for fortran
}
#endif
```

And implementation is as follows:

```
// cpp_class_method_wrapper.cpp

#include "cpp_class_method_wrapper.hh"
#include "MyCPPClass.hh"

intcpp_class_method_wrapper(int *val){
  return MyCPPClass::MyCPPClassMethod(val);
}

intcpp_class_method_wrapper_(int *val){
  return MyCPPClass::MyCPPClassMethod(val);
}
```

Remember that you are not going to be calling *object methods* directly from C or Fortran. To extract value from an existing C++ code, you will need to write the appropriate interfaces. Your C or Fortran code will call a C++ wrapper with external linkage and that wrapper will call the C++ code.

The above C++ wrapper functions are written so that they can be called from C or Fortran. Calling them from C++ (should you need to do so) requires declaring them as *extern* "C", because in this case, it signals the linker to look for functions without mangled names. Also, because these C++ functions were declared with external linkage, they must also be referenced with external linkage. This is done in the cpp_function.hh header file, included below. Observe the complete C++ program that calls all four methods:

```
// cpp_program.cpp

#include <iostream>

#include "c_function.h"
#include "cpp_function.hh"
#include "MyCPPClass.hh"

// non-CPP functions must be declared with C-style linkage
extern "C"
{
  intc_function(int *);
  intfortran_function_(int *);
}

int main(){

  int val=1;
  std::cout<< "Running CPP program with val = " <<val<<std::endl;

  // call c_function
  c_function(&val);
  std::cout<< "c_function returns val = " <<val<<std::endl;

  // call fortran_function
  fortran_function_(&val);
  std::cout<< "fortran_function returns val = " <<val<<std::endl;

  // call cpp function
  cpp_function(&val);
  std::cout<< "cpp_function returns val = " <<val<<std::endl;

  // call cpp class method
  MyCPPClass::MyCPPClassMethod(&val);
  std::cout<< "MyCPPClass::MyCPPMethod() returns val = " \
    <<val<<std::endl;

  std::cout<< "CPP program complete." << std::endl;
  std::cout<<std::endl;
  return 0;
}
```

Rather than take you through it all step by step, I am going to hand you the code and tell you what is going on. You should know enough to put it together at this point. The Fortran function looks like this:

```
! fortran_function.f90

integer function fortran_function(val)
  implicit none
```

```
  integer val
  val = val * 3
  fortran_function = 0
end function fortran_function
```

Sometimes, Fortran looks really straightforward, yes. This short code compiles to give the object file:

```
$ gfortran -c fortran_function.f90
```

which has only the Fortran-type underscore binding:

```
$ nm fortran_function.o
00000000 T fortran_function_
```

The complete C function header, with extern declarations for C++ and Fortran wrapper, looks like this:

```
// "c_function.h"

#ifdef __cplusplus
extern "C" {
#endif

intc_function(int *);

// fortran wrapper
intc_function_(int *);

#ifdef __cplusplus
}
#endif
```

And implementation is as follows:

```
#include "c_function.h"

intc_function(int *val){
  *val = *val * 2;
  return 0;
}

// fortran wrapper
intc_function_(int *val){
  return c_function(val);
}
```

The C function complies with

```
$ gcc -c c_function.c
```

and contains both the C-style (not mangled, not underscored) and Fortran style (underscored) bindings:

```
$ nm c_function.o
00000000 T c_function
00000017 T c_function_
```

The complete code for the C program, c_program.c, is as follows:

```
// c_program.c

#include <stdio.h>
#include <stdlib.h>

#include "c_function.h"
#include "cpp_function.hh"
#include "cpp_class_method_wrapper.hh"

intmain(intargc, char **argv){

  intval = strtol (argv[1], NULL, 10);
  printf("Running C program with val = %d\n", val);

  c_function(&val);
  printf("c_function returns with val = %d\n", val);

  fortran_function_(&val);
  printf("fortran_function returns with val = %d\n", val);

  cpp_function(&val);
  printf('cpp_function returns with val = %d\n", val);

  cpp_class_method_wrapper(&val);
  printf("cpp_class_method_wrapper returns val = %d\n", val);

  printf("C program complete.\n\n");
  return 0;
}
```

And the Fortran program is as follows:

```
!fortran_program.f90

program fortran_program

  implicit none
  integer, external :: c_function
  integer, external :: fortran_function
  integer, external :: cpp_function
  integer, external :: cpp_class_method_wrapper

  integer::val
  integer::return_val
  val = 1

  print *, "Running Fortran program with val = ", val
```

```
return_val = c_function(val)
  print *, "c_function returns val = ", val
  return_val = fortran_function(val)
  print *, "fortran_function returns val = ", val
  return_val = cpp_function(val)
  print *, "cpp_function returns val = ", val
  return_val = cpp_class_method_wrapper(val)
  print *, "cpp_class_method_wrapper returns val = ", val

  print *, "Fortran program complete."
  print *, ""

end program fortran_program
```

And finally, the Makefile that pulls it all together is as follows:

```
#!/usr/bin/env make

default:
    @echo "No target specified."

executables: c_program fortran_programcpp_program

run: executables
    @echo
    ./c_program
    ./fortran_program
    ./cpp_program

clean:
    rm -f *.oc_programfortran_programcpp_program

c_function.o: c_function.cc_function.h
    gcc -c c_function.c

fortran_function.o:fortran_function.f90
    gfortran -c fortran_function.f90

cpp_function.o: cpp_function.cpp
    g++ -c cpp_function.cpp

cpp_class_method_wrapper.o: cpp_class_method_wrapper.cpp
    g++ -c cpp_class_method_wrapper.cpp

MyCPPClass.o: MyCPPClass.cpp
    g++ -c MyCPPClass.cpp

function_object_files: c_function.ofortran_function.o          \
                       cpp_function.ocpp_class_method_wrapper.o \
                       MyCPPClass.o

    c_program.o: c_program.c
    gcc -c c_program.c
```

```
c_program: c_program.ofunction_object_files
    gcc c_program.oc_function.ofortran_function.ocpp_function.o \
    cpp_class_method_wrapper.oMyCPPClass.o -o c_program

fortran_program.o: fortran_program.f90
    gfortran -c fortran_program.f90

fortran_program: fortran_program.ofunction_object_files
       gfortranfortran_program.oc_function.ofortran_function.o \
         cpp_function.ocpp_class_method_wrapper.oMyCPPClass.o  \
         -o fortran_program

cpp_program.o: cpp_program.cpp
    g++ -c cpp_program.cpp

cpp_program: cpp_program.ofunction_object_files
    g++ cpp_program.oc_function.ofortran_function.ocpp_function.o \
        MyCPPClass.o -o cpp_program
```

7.6 Some Notes on Best Practices

- You now know that you can compile different parts of code with different compilers and link the results. In theory, it should also be possible to use multiple different C compilers (e.g., PGI [Portland Group], Intel, and GNU) assuming that everything is being done according to the same standards. Unfortunately, this is often not the case, and the subtleties encountered can be very, well, subtle. Linking codes from different compilers does not guarantee failure, but do not assume success. Always test.
- Mixing I/O from different languages is not recommended. I/O buffers are allocated differently, even between C and C++. (File pointers are not transferable; buffering may be different.) And even if I/O appears to work on one platform (e.g., CentOS on hardware X), do not assume that it will work the same across all platforms (e.g., Android, MacOS, or Busybox on embedded hardware).

7.7 Interoperability with Python

Python may well be the holy grail for the interoperability of scientific codes, as many popular packages do provide Python interfaces (it is actually the *primary* interface in the popular Highly Optimized Object-oriented Molecular Dynamics [HOOMD] package), and as such can be made to work together without directly addressing compatibility in the underlying codes, or at least it is a promising prospect.

Unlike C/C++ and Fortran, Python is an *interpreted* language and as such cannot be linked directly in the same way as compiled code. Being interpreted means it can be written interactively, much like a shell, but with all of the rich functionality that shell scripting typically lacks. We make the briefest of introductions here, just to show what is possible. There are several bindings available, including ctypes and simplified wrapper and interface generator (SWIG). We choose to use boost::python because it is considerably flexible.

Take this simple example of passing an array of floating-point values from Python into compiled C++ code (and, by extension, to C or Fortran) and receiving their summed value back.

Note that you will need to have Python, boost, and numpy installed.

The C++ code:

```
// in_out.cpp

// The header file contains definitions needed to generate the
// interface.

#include <boost/python.hpp>

// This function will be implemented in C++, but called from python.
float sumvals(intarray_length, boost::python::numeric::array in){
  float value = 0.0f;
  // Note that before values can be used, they must first be extracted
  // via the boost::python machinery
  for (unsigned inti = 0; i<array_length; i++)
    value += boost::python::extract<float>(in[i]);
  return value;
}

// This is where we declare what C++ functions(s) will appear in the
// python module.
BOOST_PYTHON_MODULE(in_out){
  boost::python::numeric::array::set_module_and_type("numpy", "ndarray");
    def("sumvals", sumvals);
}
```

That's it! The `def()` function is defined in boost::python and is what exposes the sumvals C++ function to the Python interpreter. That is enough to get C++ and Python talking. It gets more complicated when we begin working with objects and start handling exceptions, but you get the idea. This code compiles to the in_out.o object file. The object file gets added to a shared (.so) library. Notice that we do not link because we do not have or need a main() function. Launching a C++ code from a shell process requires main(); launching a boost::python code from a Python process replaces that paradigm.

Below is the Makefile. Remember that paths may be different on your system, so you will probably need to edit the Makefile to suit your needs.

Makefile:

```
# Where's python?

PYTHON_VERSION = 2.7
PYTHON_VERSION_NAME = py27

PYTHON_INC =/usr/include/python$(PYTHON_VERSION)
PYTHON_LIB = -lboost_python-$(PYTHON_VERSION_NAME)
```

```
# Where's Boost?

BOOST_INC =/usr/include
BOOST_LIB =/usr/lib/i386-linux-gnu

in_out.so: in_out.o
	g++ -shared -Wl,—export-dynamic in_out.o -L$(BOOST_LIB) \
	    -L$(PYTHON_LIB) -lboost_python-$(PYTHON_VERSION_NAME) \
	    -lpython$(PYTHON_VERSION) -o in_out.so

in_out.o: in_out.cpp
	g++ -g -O0 -I$(PYTHON_INC) -I$(BOOST_INC) -fPIC -c in_out.cpp

clean:
	rm -f in_out.o in_out.so
```

Finally, the Python code to drive the C++ code:

```
#!/usr/bin/env python

# We need numpy to get the python types we need.
import numpy as np

# We import our package. Python will look for in_out.so, which must
# be in the PYTHONPATH.
import in_out

vals = np.array([1.0, 2.0, 4.0], dtype=np.float64)
print(in_out.sumvals(3, vals))
```

Launch the script by either making it executable with chmod +x, launching with the 'python' command, or better yet, enter the script interactively, by hand. After all, that is the power!

For more information on using boost::python, see http://www.shocksolution.com/python-basics-tutorials-and-examples/linking-python-and-c-with-boostpython/.

8

Build Management with CMake

Ryan L. Marson and Eric Jankowski

CONTENTS

8.1 Introduction to CMake

The process of building and installing a project from source can vary in complexity from one person's computer to another, which makes it difficult to share and distribute code. CMake is a program that helps to solve the issue of cross-platform development by providing a way to build and install programs that are independent of a user's compiler, operating system, and development environment. In short, CMake automates the generation of a `Makefile` that is customized for a user's build environment, which he or she can use to build the project with `make`. In this chapter, we discuss the basics of using CMake to manage an example C++ project. First, we describe how to make and use `CMakeLists.txt`, the main input file needed by CMake to manage a project. We then show how CMake can be used to perform automated tasks, including the incorporation of version numbers into source code and the generation of customized header files. We then demonstrate how simple tests can be performed, along with system introspection that allows for conditional compilation.

8.2 `CMakeLists.txt`—Building a Basic Executable

We begin with the most fundamental aspect of compilation with CMake—the `CMakeLists.txt` file. In this example, we create `CMakeLists.txt` in a directory called `src` that will contain the source code files for the project. This file will specify how CMake should build an executable from source code files, allowing you to build your project in a manner most convenient for your workflow. The `CMakeLists.txt` file will contain all relevant information to generate native build scripts for a wide variety of platforms.

A basic CMakeLists.txt file should contain the minimum required CMake version, specifies the name of the project, and specifies the code that will be compiled and executed (Listing 8.1). In Listing 8.1, the add _ executable directive declares the name of the binary to compile and specifies the source files on which it depends, project specifics the project name, and cmake _ minimum _ required specifies the minimum acceptable CMake version.

LISTING 8.1 src/CMakeLists.txt: A Minimal Input File Used by CMake to Manage a Project

```
1 cmake _ mimimum _ required(VERSION 2.6)
2 project(Example)
3 add _ executable(Example example.cpp)
```

In this chapter, our example C++ code will double input given by the user (Listing 8.2).

LISTING 8.2 src/example.cpp: A Program That Doubles a Number

```
 1 #include<iostream>
 2 int doubleNumber(int x){return 2*x;}
 3
 4 int main(int argc,char**argv)
 5 {
 6    using namespace std;
 7    int x  =  atoi(argv[1]);
 8    cout  <<  doubleNumber(x)  <<  endl;
 9    return 0;
10 }
```

Before compiling our project, we first create a build directory within the src directory and navigate into it:

```
$ mkdir build
$ cd build
```

Building from within the build directory will avoid cluttering the src directory, making it easier to manage the files within src, facilitating the use of version control software like git or svn.

CMake uses CMakeLists.txt to create a Makefile that is customized for our build environment, which we subsequently use as input to the program make to actually build our project. To create the Makefile in the build directory we run:

```
$ cmake ../
```

after which we see that CMakeCache.txt, CMakefiles/, Makefile, and cmake _ install.cmake have been created in the build directory. The next step to building the Example executable (which we specified in CMakeLists.txt) is to invoke make, which reads the Makefile by default:

```
$ make
```

We now have a fully functional program that has been built using CMake, which we can run (with sample input/output shown) as follows:

```
$ ./Example 20
40
```

Next, we will look at how we can use CMake to create custom header files.

8.3 Generating Headers

Header files in languages like C and C++ help with code organization, permitting the declarations of variables and functions to be specified in one place that can be referred to by a number of source files. In this example, we will show how CMake can use information specified in `CMakeLists.txt` and a header template file to generate a C++ header file that will be used during project compilation. Specifically, we will show how the project version number can be inserted into a header and used in a program.

To begin, we create the file `src/ExampleConfig.h.in` (Listing 8.3), which is the header template that CMake will use to generate a header. Listing 8.3 contains one `#define` statements and looks nearly like a syntactically correct C++ header, except for the word `@VERSION_NUMBER@`. The pair of `@` symbols in Listing 8.3 wraps the name of a variable that can be set in `CMakeLists.txt` and is used by CMake to insert custom values into the header that we wish to generate.

LISTING 8.3 `src/ExampleConfig.h.in`: Header Template

```
1 // configured options and settings for Example
2 #define VERSION @VERSION_NUMBER@
```

Next, we will create a new `CMakeLists.txt` file (Listing 8.4), where lines 3–15 are added into Listing 8.1. The `set` directive on line 5 takes two arguments: the name of a CMake variable and the value to set it to, respectively. The `configure_file` directive also takes two arguments: a string specifying the path to an input file and a string specifying the path to an output file. The `configure_file` directive tells CMake to copy the input file to the output file but to substitute the variables specified in the input file. In line 9, we also see the *use* of CMake variables with the words `${PROJECT_SOURCE_DIR}` and `${PROJECT_BINARY_DIR}`: The values stored in the CMake Variables `PROJECT_SOURCE_DIR` and `PROJECT_BINARY_DIR` will be inserted into the strings on lines 9 and 10. In this example, we never specified `PROJECT_SOURCE_DIR` or `PROJECT_BINARY_DIR` explicitly; they are implicitly defined by the location in which `CMakeLists.txt` is stored and where `cmake` is invoked, respectively. The `include_directories` directive adds the paths specified by arguments to the directive to the list of directories that CMake looks into when searching for header files. In this example, we include the `PROJECT_BINARY_DIR` in the list of "include" directories because that is the directory in which we have specified our `ExampleConfig.h` header file to be created.

LISTING 8.4 src/CMakeLists.txt for Header Generation Example

```
 1 cmake_minimum_required(VERSION 2.6)
 2 project(Example)
 3
 4 #Version number
 5 set(VERSION_NUMBER 1)
 6
 7 #configure a header that contains CMake settings
 8 configure_file(
 9     "${PROJECT_SOURCE_DIR}/ExampleConfig.h.in"
10     "${PROJECT_BINARY_DIR}/ExampleConfig.h"
11     )
12
13 #add the binary tree to the search path for include files
14 include_directories("${PROJECT_BINARY_DIR}")
15
16 add_executable(Example example.cpp)
```

To summarize the example thus far: We have specified a variable that describes the version number of our project in `CMakeLists.txt`, which CMake will insert into the `ExampleConfig.h.in` template, creating `ExampleConfig.h`. The next step to using the information in `ExampleConfig.h` is to include it in `example.cpp` and reference the custom symbols (Listing 8.5). The `example.cpp` in Listing 8.5 differs from Listing 8.2 in only lines 2 and 9.

LISTING 8.5 Updated Version of Example with Version Reporting—example.cpp

```
 1 #include<iostream>
 2 #include "ExampleConfig.h"
 3 int doubleNumber(int x){return 2*x;}
 4
 5 int main(int argc,char**argv)
 6 {
 7     using namespace std;
 8     int x = atoi(argv[1]);
 9     cout << "Version: " << VERSION << endl;
10     cout << doubleNumber(x) << endl;
11     return 0;
12 }
```

We can now build and run our example as before, assuming we have navigated to the build directory:

```
$ cmake ../
$ make
$ ./Example 3
Version: 1
6
```

Looking in the build directory with ls, we see that ExampleConfig.h was created here as expected, and the make step (where your system compiler was invoked to build and link Example) proceeded without issue because we told CMake to include the build directory in the list of directories to check for header files (line 14 of Listing 8.4).

8.4 Compiling and Linking Libraries

In this section, we demonstrate how to use CMake to compile a library and link a library into our project. As an example, we will create a very simple pseudorandom number generator in src/Rand/ (Listing 8.6 and Listing 8.7).

LISTING 8.6 src/Rand/myrand.cpp: An Extremely Simple Linear Congruential Random Number Generator

```
 1 #include "myrand.h"
 2 int myrand()
 3 {
 4       static int x;
 5       x = (x*3+3)%(RAND_MAX+1);
 6       return x;
 7 }
 8
 9 void smyrand(int n)
10 {
11       for(int i=0; i<n; i++)
12           myrand();
13 }
```

LISTING 8.7 src/Rand/myrand.h: RAND_MAX Definition and Function Prototypes

```
1 #define RAND_MAX 7
2 void smyrand(int);
3 int myrand();
```

In order to tell CMake to compile myrand.cpp into a static library that can be linked against later, we need to create src/Rand/CMakeLists.txt (Listing 8.8) and add a few new directives to src/CMakeLists.txt (Listing 8.9).

LISTING 8.8 src/Rand/CMakeLists.txt: Specify the Source File in This Directory as a Dependency to a Library That Is Added to This Project

```
1 add_library(Rand myrand.cpp)
```

LISTING 8.9 src/CMakeLists.txt: Lines 16–20 and 22 Are Additions to the Previous Listing 8.4

```
1 cmake_minimum_required(VERSION 2.6)
2 project(Example)
3
4 #Version number
5 set(VERSION_NUMBER 1)
6
7 #configure a header that contains CMake settings
8 configure_file(
9     "${PROJECT_SOURCE_DIR}/ExampleConfig.h.in"
10    "${PROJECT_BINARY_DIR}/ExampleConfig.h"
11    )
12
13 #add the binary tree to the search path for include files
14 include_directories("${PROJECT_BINARY_DIR}")
15
16 #add custom random number generator
17 include_directories("${PROJECT_SOURCE_DIR}/Rand")
18 add_subdirectory(Rand)
19 set (EXTRA_LIBS ${EXTRA_LIBS} Rand)
20
21 add_executable(Example example.cpp)
22 target_link_libraries(Example ${EXTRA_LIBS})
```

As with line 14, the `include_directories` on line 17 tells CMake to include `src/Rand` to the list of directories to look for header files in. The `add_subdirectory` directive on line 18 tells CMake to expect a `CMakeLists.txt` file in `src/Rand` and to parse it. The `set` directive on line 19 appends `Rand` to the `EXTRA_LIBS` variable, which is used in line 22 to specify which libraries should be linked against when creating the `Example` executable.

In order to use the linked random number generator in our example, we modify `src/example.cpp` to create Listing 8.10.

LISTING 8.10 src/example.cpp: Seed the Random Number Generator with the Argument Passed into main() and Print Out Two Periods of Random Numbers

```
1 #include<iostream>
2 #include "ExampleConfig.h"
3 #include "Rand/myrand.h"
4 int doubleNumber(int x){return 2*x;}
5
6 int main(int argc,char**argv)
7 {
8     using namespace std;
9     int x = atoi(argv[1]);
10    cout << "Version: " << VERSION << endl;
11    cout << doubleNumber(x) << endl;
```

```
12        smyrand(x);
13        for(int i=0;i<14;i++)
14            cout << myrand() << " ";
15        cout << endl;
16        return 0;
17 }
```

Now, after navigating to `src/build`, we can have CMake and make compile and link our project, after which we can run it:

```
$ cmake ../
$ make
$ ./Example 3
Version: 1
6
0 3 4 7 0 3 4 7 0 3 4 7 0 3
```

Inspecting the `src/build/Rand` directory, we find that `src/build/Rand/libRand.a` has been compiled, and based on the successful execution of `Example`, it appears that our executable was successfully linked against `libRand.a`. Here, CMake has been quite handy in creating Makefiles that automatically named and compiled object files, compiled a library, and linked a custom library into our example executable!

8.5 Testing

Another handy feature of CMake is the ability to create a suite of tests that can then be executed by CTest, a testing tool that is distributed with CMake. In this section, we will show how to write a test that checks the output of a program against an expected result. Thus far, we have created an example project (Listing 8.10) that reads in an argument from the command line, converts it to an integer, prints out the project version number, doubles the user's input, seeds a (quite bad) random number generator, and then prints out a few iterations of the random number generator. The output at the end of the previous section, `0 3 4 7 0 3 4 7 0 3 4 7 0 3`, looks suspicious. Specifically, we expect our random number generator to cycle through each number from 0 through `RAND_MAX` once before repeating. Each cycle through the eight digits of the random number generator is called one period. We also know, by working through the math on line 5 of Listing 8.6, that if we seed the generator with `x=0`, the first number returned by `myrand()` should be 3. In fact, we actually expected the period of our random number generator to be `3 2 5 4 7 6 1 0` when seeded with 0, by virtue of working out an example on paper.

Using our expected output, we can use CMake to create and run a test for that output. To begin making our new test, we add three new directives (lines 24–30) to the end of `src/CMakeLists.txt` (Listing 8.11). The `include` statement enables the CTest directives that follow to be used. The `add_test` directive declares a new test called `CheckSeed` that will be invoked by passing the argument `0` to the `Example` executable. Lines 26–30 specify the criteria for passing the `CheckSeed` test. Specifically, this `set_tests_properties` directive tells CTest to search for the string `3 2 5 4 7 6 1 0` in the output of `./Example 0`. If the search string is found, the test will register as being passed, and fail otherwise.

LISTING 8.11 **`src/CMakeLists.txt`**

```
1 cmake_minimum_required(VERSION 2.6)
2 project(Example)
3
4 #Version number
5 set(VERSION_NUMBER 1)
6
7 #configure a header that contains CMake settings
8 configure_file(
9       "${PROJECT_SOURCE_DIR}/ExampleConfig.h.in"
10      "${PROJECT_BINARY_DIR}/ExampleConfig.h"
11      )
12
13 #add the binary tree to the search path for include files
14 include_directories("${PROJECT_BINARY_DIR}")
15
16 #add custom random number generator
17 include_directories("${PROJECT_SOURCE_DIR}/Rand")
18 add_subdirectory(Rand)
19 set (EXTRA_LIBS ${EXTRA_LIBS} Rand)
20
21 add_executable(Example example.cpp)
22 target_link_libraries(Example ${EXTRA_LIBS})
23
24 include(CTest)
25 add_test(CheckSeed Example 0)
26 set_tests_properties(CheckSeed
27      PROPERTIES
28      PASS_REGULAR_EXPRESSION
29      "3 2 5 4 7 6 1 0"
30      )
```

To build and invoke our tests, we navigate to `src/build` and run:

```
$ cmake ../
$ make
$ ctest
```

after which we see output from CTest indicating that our test has failed. We can inspect the details of our test by printing out the results of the last test:

```
$ cat Testing/Temporary/LastTest.log
```

which gives something like Listing 8.12.

LISTING 8.12 src/build/Testing/Temporary/LastTest.log

```
 1 Start testing: Nov 19 20:38 MST
 2 ----------------------------------------------------------

 3 1/1 Testing: CheckSeed
 4 1/1 Test: CheckSeed
 5 Command: "/Users/erjank/src/build/Example" "0"
 6 Directory: /Users/erjank/src/build
 7 "CheckSeed" start time: Nov 19 20:38 MST
 8 Output:
 9 ----------------------------------------------------------
10 Version: 1
11 0
12 3 4 7 0 3 4 7 0 3 4 7 0 3 4
13 <end of output>
14 Test time =   0.02 sec
15 ----------------------------------------------------------
16 Test Fail Reason:
17 Required regular expression not found.Regex=[3 2 5 4 7 6 1
18 0]
19 "CheckSeed" end time: Nov 19 20:38 MST
20 "CheckSeed" time elapsed: 00:00:00
21 ----------------------------------------------------------
22
23 End testing: Nov 19 20:38 MST
```

Now that we have seen the confusing output a few times, we hypothesize that we have made a typo in `myrand()`. Let's see what happens with the multiplier of our random number generator changed to 5 (Listing 8.13).

LISTING 8.13 src/Rand/myrand.cpp: Line 5 Updated with 5 as the Multiplier Instead of 3

```
 1 #include "myrand.h"
 2 int myrand()
 3 {
 4     static int x;
 5     x = (x*5+3)%(RAND_MAX+1);
 6     return x;
 7 }
 8
 9 void smyrand(int n)
10 {
11     for(int i=0; i<n; i++)
12         myrand();
13 }
```

Now, we can recompile our code and rerun the tests as simply as:

```
$ make
$ ctest
Test project /Users/erjank/src/build
    Start 1: CheckSeed
1/1 Test #1: CheckSeed .......................
    Passed    0.00 sec

100% tests passed, 0 tests failed out of 1
```

This time, it looks like with 5 as the multiplier in our random number generator, we get the expected output!

8.6 Conditional Compilation and System Introspection

The final CMake feature we will cover in this introduction will permit us to investigate code provided by a user's system to determine how a project is built. In our example thus far, we have implemented an extremely simplistic pseudorandom number generator that has a very short period of just eight numbers. Our random number generator should not be used for any serious computation, but it might be appropriate to use in a pinch, if no other random number generator is defined on a system. In this section, we will show how to use CMake to test for the existence of a random number generator, use the system random number generator if it exists, and use our random number generator otherwise.

To begin, we modify `src/CMakeLists.txt` by adding lines 4 and 5 in Listing 8.14

LISTING 8.14 `src/CMakeLists.txt`

```
1 cmake_minimum_required(VERSION 2.6)
2 project(Example)
3
4 include (${CMAKE_ROOT}/Modules/CheckFunctionExists.cmake)
5 check_function_exists (rand RAND_EXISTS)
6
7 #Version number
8 set(VERSION_NUMBER 1)
9
10 #configure a header that contains CMake settings
11 configure_file(
12     "${PROJECT_SOURCE_DIR}/ExampleConfig.h.in"
13     "${PROJECT_BINARY_DIR}/ExampleConfig.h"
14     )
15
16 #add the binary tree to the search path for include files
17 include_directories("${PROJECT_BINARY_DIR}")
18
```

```
19 # our custom random number generator
20 include_directories("${PROJECT_SOURCE_DIR}/Rand")
21 add_subdirectory(Rand)
22 set (EXTRA_LIBS ${EXTRA_LIBS} Rand)
23
24 add_executable(Example example.cpp)
25 target_link_libraries(Example ${EXTRA_LIBS})
26
27 include(CTest)
28 add_test(CheckSeed Example 0)
29 set_tests_properties(CheckSeed
30     PROPERTIES
31     PASS_REGULAR_EXPRESSION
32     "3 2 5 4 7 6 1 0"
33     )
```

which permits the `check_function_exists` directive to be called, checking for `rand` in the linkable system libraries. It is important that the `check_function_exists` call comes before the `configure_file` directive that manipulates `build/ExampleConfig.h.in` because we wish to use `RAND_EXISTS` within `build/ExampleConfig.h.in`. If `rand` is found in a system library, then the `RAND_EXISTS` variable is defined, which we make use of in `ExampleConfig.h.in` on line 2 (Listing 8.15).

LISTING 8.15 `src/ExampleConfig.h.in`

```
1 #define VERSION @VERSION_NUMBER@
2 #cmakedefine RAND_EXISTS
```

Simply by including line 2, we find that `build/ExampleConfig.h` contains `#define RAND_EXISTS` after it is generated by CMake, but only if the `rand` function is found to be linkable by CMake. We next modify `src/Rand/myrand.h` (Listing 8.16) and `src/Rand/myrand.cpp` (Listing 8.17) to use the `RAND_EXISTS` preprocessor macro to determine if our simple random number generator should be used, or if the system's `rand` should be used:

LISTING 8.16 `src/Rand/myrand.h`

```
1 #ifndef RAND_EXISTS
2 #define RAND_MAX 7
3 #endif
4 void smyrand(int);
5 int myrand();
```

LISTING 8.17 `src/Rand/myrand.cpp`

```
1 #include "ExampleConfig.h"
2 #ifdef RAND_EXISTS
```

```
 3 #include <cstdlib>
 4 #endif
 5 #include "myrand.h"
 6 int myrand()
 7 {
 8     #ifdef RAND_EXISTS
 9     return rand();
10     #else
11     static int x;
12     x = (x*5+3)%(RAND_MAX+1);
13     return x;
14     #endif
15 }
16
17 void smyrand(int n)
18 {
19     #ifdef RAND_EXISTS
20     return srand(n);
21     #else
22     for(int i=0; i<n; i++)
23         myrand();
24     #endif
25 }
```

We invoke CMake from the src/build directory again to create an updated src/build/ExampleConfig.h as well as the necessary makefiles:

```
$ cmake ../
```

After recompiling and running:

```
$ make
$ ./Example 0
Version: 1
0
520932930 28925691 822784415 890459872
145532761 2132723841 1040043610 1643550337
68362598 66433441 2002830094 1906706780
1269870926 1028169396
```

we find that the numbers returned by our Example executable are now much larger than before. Here, the larger, different numbers are a result of the system rand being used instead of our custom myrand.

8.7 Summary

In this chapter, we started with a minimal `CMakeLists.txt` file and added features successively to demonstrate some basic ways that CMake can help with software development. The features explained here are only a small subset of the capabilities available in CMake but should provide a new user with enough of a foundation to drastically reduce the work associated with building projects for multiple environments. Consider using CMake to automatically generate makefiles in projects that require frequent editing of these files and for automating the creation and running of unit tests for ensuring the project's correctness.

9

Getting Started with Python 3

Brian C. Barnes and Michael S. Sellers

CONTENTS

9.1 Python 3 Essentials

This section outlines several important aspects of Python 3 and aims to soften the transition to Python from languages such as C/C++ or Fortran. Similar to these other languages, we discuss *types, statements, control flow,* and *object oriented* functionality in the context of programming for scientific and technical purposes. Python also possesses useful components such as *modules* that greatly aid the scientific programmer. With an understanding of these essentials, inexperienced programmers should be well on their way to writing their first Python script, while experienced programmers should feel comfortable diving right in to the language and beginning to create useful code.

9.1.1 Running a Python Script

Python is a runtime compiled language, like Java and unlike C/C++ and Fortran. This means that to execute a Python script, the user must invoke the Python interpreter to parse, compile, and execute. This is all done in one invocation and usually performed in two ways. The first is by invoking the interpreter from the command line and passing the script, in this example, called `my_script.py`.

```
[user@host ~]$ python3 my_script.py
```

The second is by including the following line at the top of `my_script.py` to tell the command shell to call the interpreter:

```
#!/usr/bin/env python3
```

If the second method is chosen, the user can run the script file directly and should ensure it has executable permissions.

```
[user@host ~]$ chmod u+x my_script.py
[user@host ~]$ ./my_script.py
```

9.1.2 Built-In Types and Their Uses

Below, we visit some of the most useful and common Built-in types in Python. These will become the building blocks of your Python script or program, and most have analogs in C/C++, Java, and Fortran. What sets Python apart is that many of these common types have functions within them that take the place of commonly written code. The user also does not have to declare any of these common types, as would occur in C/C++ or Java—just define them to start using in your code. Once we have learned a few of the essential Built-in types, we will see how they can be used with *statements* and *control flow* to build more complex scripts or programs.

9.1.2.1 Lists

Lists are Python's containers that hold objects in a given order. These might sound like arrays to C/C++ or Fortran followers; however, they are slightly different. The "array" in Python does exist, with the addition of an external module, and will be covered later in this chapter. Experienced users of C++ will recognize a similar list container from the Standard Library.

Python lists possess several Built-in functions to operate on themselves, which can alleviate some of your coding burden. Creation and access of a list in your script are simple using square brackets, [], with an index beginning at 0 (Python lists are "zero-indexed"). Lists can span multiple dimensions as well, by creating nested lists. Keep in mind, if you create multiple lists and then create a nested list from these, your nested list will contain pointers to your original lists. Any change in the original lists will change your nested list. You can check the length of a list using the `len(list)` function. The example below illustrates many of these points. In examples meant to be run at an interactive prompt, we will use the symbols >>> to denote the interpreter and any lines without those symbols represent output.

```
>>>dates = []                    # create a list called dates
>>>dates.append(1991)            # add an element with integer 1991
>>>dates.append(1992)
>>>dates.append(1993)
>>>dates.append(1992)
print(dates)
[1991,1992,1993,1992]
>>>print(dates[1])
1992
>>>print(len(dates))
4
>>>dates.remove(1992)            # remove first element containing 1992
print(dates)
[1991,1993,1992]
>>>sixties = [1960, 1961, 1962]       # create and fill sixties
>>>seventies = [1970, 1971, 1972]
>>>eighties = [1980, 1981, 1982]

# set the zeroth element in dates equal to the sixties list
>>>dates[0] = sixties
>>>dates[1] = seventies
>>>dates[2] = eighties

>>>print(dates)
[[1960, 1961, 1962], [1970, 1971, 1972], [1980, 1981, 1982]]
>>>print(dates[1][2])
1972
>>>sixties.remove(1961)          # remove element from a nested list
>>>print(dates[0])
[1960, 1962]
```

A fast way to specify and access certain portions of a list is through a technique called *slicing*. Slicing is a way of passing multiple indices to a list using a colon to denote a range, to get back a smaller list. The first index *a* is inclusive, while the second index *b* is exclusive; this is written mathematically as [*a*:*b*). If you want a range starting from the beginning or finishing at the end, you do not have to use a first or second index, respectively. For example, if we have a list `pizza` of length 9, we can get the first four elements—[0], [1], [2], and [3]—using `pizza[:4]` (which is the same as `pizza[0:4]`). Similarly, with the last four elements, we use `pizza[5:]`. If we just want the middle 3, we call `pizza[3:6]`, where the slice stops before index 6. Python allows for negative numbers in list indices as well, where `pizza[-1]` will give you the last entry and `pizza[:-1]` returns everything but the last entry. Accordingly, `pizza[-2]` will give you the second to last entry.

9.1.2.2 Tuples

Tuples are another type of container for objects, but they are *immutable*, or unchangeable. Once created, their contents cannot be altered. They are made just like lists using the parentheses, `()`, and for a tuple called `tnames =('my','name','is')` or `t123 = (1.0,2.0,3.0)`, setting `tnames[0] = 'your'` or `t123[2] = 4.0` would return an error.

9.1.2.3 Strings

Strings are objects that handle text within Python and they are also the basic unit, or character, of text data. Creating a string is simple in Python using single or double quotations, and printing the string is just as simple.

```
name = "John Cleese"        # create a string
print(name)                 # John Cleese
```

The string object `name` is actually a list of *immutable* strings, each with a length of 1. So, printing `name[0:4]` would return `John` and setting `name[0:4] = 'Brad'` would return an error. If you would like to change `John` to `Brad`, one would need to create another string, such as `'Brad'+name[4:]`. There are also many Built-in string functions. For example, to split a line of text into a list of string objects, one could use `''.split(name)`, or `name.split()` to operate on `name` and remove whitespace. To join an array of string objects into a single string object, use `''.join(name)` or `name.join()`. Below is an example of this process.

```
>>>name = "John Cleese"                 # create a string
>>>firstlast_name = name.split()        # split string into a list
>>>print(firstlast_name)
['John', 'Cleese']
>>>print(firstlast_name[0])
John
>>>name2 = ''.join(firstlast_name)      # combine list of strings
>>>print(name2)
JohnCleese
```

9.1.2.4 Dictionaries

A *dictionary* in Python is a type not directly available in Fortran, and similar to the *map* container in the C++ Standard Library. It is useful if you want a container for your data and are able to access the data by a label. They consist of key/value pairs separated by a colon in an array and are unsorted. Dictionaries are constructed similar to lists, but their syntax uses curly brackets, {}. The indices of a dictionary are the keys and can be any *immutable* object, so for a given dictionary `monte = {'spam':1.0, 'parrot':0.0}`, the keys would be `spam` and `parrot`. To access an item in a dictionary, just use its key as the index: `monte['spam']` returns `1.0`. The function `.keys()` operating on a dictionary will return a list of the dictionary's keys.

9.1.3 Statements and Control Flow

Python controls the decisions that a computer can make in a similar way as other languages, through `if/then/else` statements and loop structures with specified termination criteria. What is quite different though is the formatting of these structures and their syntax. First, where C++ and Java use {}, Python uses indenting. Levels of control flow are determined by the indentation level of your code. Four spaces are commonly used, but that is not a rule. In general, lines with the same indentation level are considered to be grouped for purposes of control flow. More on this and *Pythonic* style in Section 9.3.1. Second, while conditional statements are similar to C/C++, Fortran, and Java, the syntax of `if` statements and `for` and `while` loops and how they iterate are different. Let's look at some examples.

9.1.3.1 *if/then/else Statements*

`if` statements use standard comparison expressions such as `<, >, <=, >=, ==`, and `!=` to test a truth value. They also incorporate more natural language such as `is`, `is not`, `and`, and `or`. If you want to test whether an element is in a list, you can use the expression, `if` *`item`* `in` *`list`*, where `item` and `list` are the value and list name you wish to test.

```
color = "green"
if color is "blue":
    print(color,"is my favorite!")
elif color is "green":
    print(color, "is my second favorite.")
elif color is not "red" and color is not "purple":
    print("At least it's not red or purple!")
else:
    print(color,"is okay, but it's not the best.")
```

9.1.3.2 *Looping*

Python has some useful functions for generating a list of items to loop over. These *generators* are included in the loop structure and iterated over. A common command for this is `range()`. This will return a sequence of numbers in a desired span, such as `range(10)` for numbers 0 through 9. Passing two integers to range, `range(4,50)`, will produce a sequence beginning with the first number and a length according to the second. For a sequence of nonconsecutive numbers, say every two, use `range(4,50,2)`. Below is an example using `range(5)`.

```
years = [2001, 2002, 2003, 2004, 2005]

for i in range(len(years)):
    # A for loop over a sequence from the range function.
    # "i" is a number in the sequence.
    print("Year #", i, "is", years[i])
```

Two other useful sequence producers are `enumerate()` and `zip()`. `enumerate()` takes an iterable argument, such as our `eighties` list from a previous example, and returns a list of tuples where the first element is a counter and the second is an item in the list. For example, `enumerate(eighties)` would return `[(0, 1981),(1, 1982),(2, 1983)]`. It is a convenient way to access both the index and list element for use in a loop. `zip()`takes multiple iterable arguments and returns tuples of the elements with equal indices. So, `zip(seventies,eighties)` would give `[(1970, 1980), (1971, 1981), (1972, 1982)]`. This is useful for looping over multiple lists at the same time.

Finally, if you only want to loop over the elements in a list, you can use a simple for loop construction, `for` *`item`* `in` *`list`*`:` to loop over the elements in `list`, set to `item` each step. This is shown in the example below using the list ages.

```
ages = [10, 20, 30, 40, 50]
cy = 2015
print("The year is", cy)
for age in ages:
# For loop over the elements in the ages list.
# "age" is the element of ages.
    print("Age:", age, "Born:", cy-age)
    for j in range(0, 10, 2):
        print("In", cy+j, "age is: ", age+j)

while cy < 2020:
    cy = cy + 1
    print("The year is now", cy)
```

9.1.4 Functions

A *function* in Python is a way of writing reusable code similar to that found in many other programming languages. Function usage in Python is similar to their counterparts in C/C++ and Fortran but naturally has a unique syntax.

```
def name_of_function(parameter):
    # "def" is required to denote the declaration of a function
    parameter = parameter + 1
    return parameter

age = 10
newage = name_of_function(age)
# Create a variable age and pass it to our function.
print(newage) # should return '11'
```

Functions can return any type of object using `return` or will return `None` if `return` is not specified (`None` is Python's `NULL` equivalent). One important distinction between Python functions and other languages is the *call by* style. C/C++ and Fortran users are familiar with *call by reference*; however, Python behaves differently when it comes to immutable objects. This is shown in our example above. In a call by reference style, we would not need to return `parameter`—simply adding 1 to `parameter` would change the variable we passed in, age, to `11`. However in Python, *immutable* objects cannot be changed "in-place." Lists, dictionaries, or other mutable objects are allowed to be changed within the function.

Python functions may also set variable defaults. If an argument is not passed for this parameter, then it retains its value set in the function declaration. See the example below.

```
def pet_shop(animal, alive=True):
    # Define a function with the alive variable set to a default of True.
    print("Hello?!")
    if animal == "parrot":
```

```
        if not alive:
            print("Refunds in Bolton!")
        else:
            print("What a lovely parrot!")

my_pet = "parrot"
pet_shop(my_pet)
        # Hello?!
        # What a lovely parrot!
pet_shop(my_pet, False)
        # Hello?!
        # Refunds in Bolton!
```

9.1.5 Modules

One of Python's best traits is that it carries around a mountain of prewritten code in containers called *modules*. This means that if you are coding in Python, there is a good chance that some of the programming work is already done for you! Whether it is converting between strings and binary data, generating random numbers, managing threads, interacting with a web server, or zipping files, all of this code is waiting within the Python source. Let's now look at how to access it with some descriptions and examples of common *modules*.

9.1.5.1 The sys Module

`sys` is a commonly used module and provides some basic ways to interact with the Python interpreter. To use `sys`, simply add the line `import sys` to your script before you write code to access the `sys` commands. When you issue the `import` command, Python will read a file within its source and its contents will become the `sys` object. Any functions within the sys module file can now be accessed within your script as `sys.a_func()`, where `a_func` is the name of the desired function. Just a tip, when functions exist within other objects, they are referred to as *methods*. Two useful objects or methods within the `sys` module are `argv` and `exit()`. The first is a list of command line inputs to the Python interpreter. The first element in the list is the Python script filename. The second method, `exit()`, will terminate the Python program and interpreter. The method's parameter defaults to `0`, but any nonzero integer can be provided to return an exit status.

```
# contents of myscript.py
import sys

print("Hello World!")
print("My name is", sys.argv[1]) # access argv list in sys module
print("My script's name is",sys.argv[0])

# at the command line...
[user@host ~]$ python myscript.py John
        Hello World!
        My name is John
        My script's name is myscript.py
```

9.1.5.2 The io Module

Another useful module is `io`. This makes reading in any type of file quite easy. Use the `io.open()` method get a handle on a file and open the stream, `io.readline()` or `io.readlines(lines)` to read in a line or multiple lines of a file, and `io.close()` to close the stream to the file. Within the `.open()` method, you can specify many arguments to describe the file name to open, the type of file, and whether you would like to read (r), write (w), or append (a) to it. Let's look at an example using an empty text file, `parrots.txt`, to which we add three lines of text.

You can also use the statement `with io.open('filename','w') as my_file:`, and once the code within this statement is executed, the file stream will close. Note that `open()` is a language Built-in function; if that is the only `io` function of interest to your script, you do not need to import `io`.

```
import io

with io.open("parrots.txt","a") as my_file:
# Alternative way to open and append.
# Notice the "\n" to create a newline.
    my_file2.write(u"Polly\n")          # the u prefix ensures unicode
    my_file2.write(u"Piney\n")
    my_file2.write(u"Fjordy\n")

names = []
my_file = io.open("parrots.txt","r")   # open parrots.txt for reading
names.append(my_file.readline())       # read the first line
names.append(my_file.readline())       # read the second line
my_file.close()

for i, item in enumerate(names):
    print("Parrot #", i, "is", item)   # print read-in lines from the list
```

9.1.5.3 Importing Modules

When importing a module, one can use various syntaxes to achieve different naming conventions. Using `import` *`module`* `as` *`custom_name`* will allow you to name the module object for use in your script. As an example, if we call `import sys as foo`, we can now access `argv` using `foo.argv`. This is handy when you want to carefully control the namespace within your script, especially if you have variables or functions that share names with your imported module. Specific functions may be imported from a module with the following syntax: `from` *`module`* `import` *`function_name`*. This is often used for importing from the `math` package, and will be used in a later example.

9.1.6 Other Essential Modules

`os`	Provides a portable interface to operating system functionality
`re`	A regular expressions module, providing functionality similar to Bash or Perl for regex
`time`	Useful for clock-based operations such as profiling or sleeping
`collections`	Provides several container datatypes that are useful alternatives to Python's Built-in types
`itertools`	Contains many different iterators that accept a variety of inputs, including combinatoric generators
`random`	Pseudorandom number generator functionality supporting various distributions
`argparse`	An easy way to add command-line interface functionality to Python scripts
`logging`	Tools to log messages at several different warning levels and can replace print statement based debugging

9.1.7 A Python Class

Python is an object-oriented (OO) language, which means that everything is an object. Objects hold information, have defined *types*, and can perform operations on themselves or other objects. If that scares you, it shouldn't! Most Python scripts are short and you will not necessarily have to code in the OO world of *methods, classes,* and *inheritance*. These are most useful for larger program development. Plus, if you have made it this far, you have been working with objects already! One of the features that makes Python so useful, the *module,* is highly object oriented, but as we just learned, its use is quite straightforward, and its OO nature is unobtrusive. Essentially, with Python, you are always doing a small amount of object-oriented programming. We continue with a summary of the Python class and how it can be useful in scientific programming.

The *class* is an object for partitioning code that you will reuse many times, or code you will use multiple copies of at the same time. It is a container for variables and functions that represent a common theme, and these can be accessed in our Python script, just as we access functions in a *module*. They can exist in our script or in separate files just like a module. The difference is, we can make multiple *instances* (copies) of the same class, change the value of variables in specific instances, and move these instances around as elements in lists. In short, classes are an abstract way of organizing data, and the copies we create allow us to make productive use of this organization.

Consider this programming case: we would like to represent 10 points on a two-dimensional (2D) grid. This is simple enough to do with lists of length 2 in a list of length 10 or by creating a list of tuples. If we want to move these points, that is, change their value in the list, we could do so uniformly with a `for` loop and by adding 1.0 to each element. If we would like our system to be more complex, for example, each point has a color, a diameter, and the ability to determine how far away from an origin our points were, our list could quickly grow to a size that is difficult to manage and access within the code.

Instead of keeping track of all the list elements in our point list, we can create a `GridPoint` class to hold all this information, with variables and lists that are intuitively named and easy to access. Functions in the `GridPoint` class must have their first argument be `self`. Self is like `this` in C++ or Java and refers to the class whose method was called; however, when calling a class' methods you do not need to include `self`. Take a look at the example below, where we create a `GridPoint` class in a separate file named `grid.py`, and then in our script, we make multiple instances of the class to represent each point. Each copy of the class will contain variables of the same name but will be separate from each other in the Python namespace.

```
# Separate file called grid.py
from math import sqrt

class GridPoint(object):
    def __init__(self):
    # Executed upon the creation of the class.
        self.color = None
        self.diameter = None
        self.pts = []
        self.pts.append(0.0)
        self.pts.append(0.0)

    def magnitude(self):
    # Requires no arguments when called, must define with "self"
```

```
        mag = self.pts[0]*self.pts[0]+self.pts[1]*self.pts[1]
        return sqrt(mag)

# ---------------------------------------------------------------------------
# ---------------------------------------------------------------------------

# Our script file.
from grid import GridPoint

# Create instance of the GridPoint class called mypoint, set list elements.
mypoint = GridPoint()
mypoint.pts[0] = 1.0
mypoint.pts[1] = 2.0

print(mypoint.magnitude())          # 2.23606797749979

grid = []                           # Create a list.
for gpoint in range(10):            # For loop over a sequence.
    # Create an instance of the GridPoint class and add it to the grid list.
    grid.append(GridPoint())
    # Set some variables within each class.
    grid[gpoint].color = "blue"
    grid[gpoint].diameter = 5.0
    grid[gpoint].pts = [1.0,1.0]
```

9.2 Python Number Crunching

9.2.1 How Python Handles Numbers

Python has two commonly used types of numbers: `int` and `float`. This is similar to other languages, such as C, except in Python, every float is stored with a double-precision IEEE 754 representation. All the caveats regarding IEEE 754 Floating-Point Arithmetic in other languages also apply to Python. Basic operators include +, −, *, /, = (assignment), `**` (power), `%` (modulo), and // (floor division). Common mathematical functions are available in the math module of the standard library. To print a floating-point representation of π, and then print its cosine, one performs the following:

```
import math
print(math.pi)
print(math.cos(math.pi))
```

If you are frequently using common mathematical functions, the repeated use of the `math` prefix can make source code difficult to read. Therefore, `math` is one of the few modules for which it is generally acceptable coding style to perform a "`from module import *`" instead of "`import module`". This chapter generally follows that convention.

When a math operation involves operands of a different type, Python automatically converts the integer to a float before performing the operation. It will not convert the actual object, but just the mathematical value used during the operation. Further, in Python 3, division converts its operands to float and returns float. This is different from the multiplication operator, which will return an int if its operands are both of type int. Converting an int to a float is straightforward: the float() Built-in function will return a floating-point number given a number or string. Converting a float to an int is similarly direct using int() if you wish to truncate downward, but other rounding operations are available: the Built-in round() to get the nearest integer, or ceil() in the math module to round upward. Let's try three different ways to calculate the distance between two points in a plane. For clarity, we will assume that you are running this in an interpreter or will add your own print() functions.

```
from math import *
class Point:
    x = 0.0
    y = 0.0
a = Point() ; b = Point()
b.x = 5.0 ; b.y = 5.0
sqrt((b.x-a.x)**2+(b.y-a.y)**2)                # option 1

a = [0,0] ; b = [5,5]
sqrt((b[0]-a[0])**2+(b[1]-a[1])**2)            # option 2
sqrt(sum([(b[i]-a[i])**2 for i in [0,1]]))     # option 3
```

The first part of the example displays an object oriented approach to finding the distance between two points, a and b. A class is defined, that has attributes x and y with default values of 0.0. Two objects are instantiated, a and b, with a retaining the default values and b having its attributes assigned new values. The Euclidean distance is calculated using the x and y coordinates of each point.

Programmers from different backgrounds, such as Fortran authors, or those who wish to have their code look more like vector math may prefer the latter half of the example. In that section, points are defined as Python lists, a and b, of what are assumed to be x and y coordinates. The distance calculation is performed by accessing list indices instead of Point object attributes. In the final square-root calculation, the Built-in function sum and a "list comprehension" are used. This generalizes the iteration over list indices instead of having to explicitly write out the summation for each element.

A list comprehension is enclosed in brackets, similar to writing a list, and contains an expression followed by a for clause (and possibly optional clauses). The list that is created is the result of evaluating the expression on the objects generated by the for clause. It is a more *Pythonic* way to do things than the second square-root calculation. All three square-root calculations should yield identical results of $5\sqrt{2}$, within IEEE 754 machine precision.

9.2.2 Numpy

If you perform calculations on lists of numbers frequently, or want to do common vector or matrix operations, the numpy module is quite useful. It is not part of the standard library, so you will need to install it or use a Python distribution (such as Anaconda, discussed in Section 9.3.5) that includes it by default. The fundamentally important object type that numpy introduces is ndarray, a multidimensional array object. There are also a great many routines for performing a variety of mathematical operations on those arrays. As an object to be used in numerical computation, `ndarray` is preferable to `list`. Operations involving numpy arrays are much faster than numerical work on lists. Fortunately, it is also easy to convert a `list` to `ndarray`. Let's look again at the distance calculation, now using `numpy`:

```
import numpy as np
a = np.array([0,0]) ; b = np.array([5,5])
np.linalg.norm(b-a)
```

By casting our points as numpy arrays, we are able to make use of two particular numpy features to make this calculation simpler: an array operator for subtraction, and numpy's `norm` function from its linear algebra library. The Euclidean distance is the L^2 norm.

It is worth taking a moment to review some important features of numpy, as the module is quite extensive and we will be unable to go over all its functionality in this text. Also within the linear algebra library is `dot()`, which takes two arrays as input and computes their dot product. When the arrays are rank 2, they are matrices, and `dot()` is the matrix multiplication operation. General categories of linear algebra available include array product methods, decompositions, eigenvalue solvers, and array inverse operations. Most of numpy's numerical routines are capable of handling fully N-dimensional arrays. However, often, a scientist will be working with time series data for one variable, which is handled in numpy as a one-dimensional (1D) array (vector). Let's do an example with the polynomial module:

```
import numpy as np
data = np.array([x**3+np.random.rand() for x in np.arange(10)])
np.polyfit(np.arange(len(data)), data, 3)
```

Above, we create a 1D array object using the numpy functions `rand()` and `arange()` in a list comprehension. The `arange()` function simply returns evenly spaced values as a numpy array, much like the Python `range()` function. The resulting vector is similar to: `[ 0   1   8  27  64 125 216 343 512 729]`, but with random noise in the range of 0 to 1 added to each value. Note that when you print a numpy array object, there are no commas between elements! This is one way it is visually different from a Python list. After generating the "data" vector, we use the numpy `polyfit` function to fit the data to a third-degree polynomial. Naturally, this should fit quite well, but not perfectly. The function `polyfit` returns a 1D array of optimal polynomial coefficients for the data, with the highest order term first. It will be close to one, and the others closer to zero. It is common when writing a script to assign the results of polyfit to an object for reuse elsewhere, but in an interpreter, the examples as written will output to screen. The polynomial package contains a great number of useful functions for fitting, calculus, and algebra. It applies not only to common polynomials but also to special classes such as Chebyshev, Legendre, Laguerre, and Hermite polynomials.

Essentially, numpy lays the groundwork for numerical computation using arrays. Although not featured here, its basic functions of array manipulation for reshaping, sorting, masking, transpose, and related operations are useful operations not available for a Python list. Numpy also has a robust C-types foreign function interface for creating objects from calculations originally run in C routines. For advanced numerical computing, most development is now occurring in modules such as scipy that import numpy to make use of its ndarray objects and methods.

9.2.3 Scipy

Scipy is Python's counterpart to the venerable "Numerical Recipes" textbook (in C or Fortran) by Press, Teukolsky, Vetterling, and Flannery. It contains subpackages for special functions (e.g., Bessel), integration, optimization, interpolation, Fourier transforms, signal processing, linear algebra, eigenvalues, mathematical graphs, statistics, and multidimensional image processing. There is some overlap in functionality with numpy, for example, the linear algebra package, and in general, the scipy implementations will be either more full featured or more recently updated. However, it is good practice to import the fewest number of packages necessary for your Python program. If numpy's linear algebra package is solving your problem correctly and efficiently, there is no need to import scipy. Sometimes, a scipy function is simply an interface to call the identical numpy function. Let's look at a quick example using scipy's `optimize` module, as optimization is a problem ubiquitous in science and engineering:

```
from scipy.optimize import minimize
def func(x):
    return x**3-x**2

minimize(func, 0.1)                     # option 1
minimize(lambda x: x**3-x**2, 0.1)     # option 2
```

The above example optimizes the function $f(x)=x^3-x^2$ to a local minimum, using an initial guess of 0.1. The algorithm used is quasi-Newton "BFGS" (named after its authors' initials), although that can be changed by specifying an optional method parameter. The result returned finds the minimum of 2/3, to within the default tolerance. For this author, the result for x was `0.66666539`. Note that the two `minimize` calls are equivalent and return the same result. The second option is included in order to demonstrate use of a `lambda` function, which is a way to define functions for one-time use "on the fly" instead of in a separate block of code. Lambdas are part of standard Python but are sometimes difficult to interpret and should be used sparingly. The optimize package has a large variety of optimization methods implemented, not only for local minimization but also for global or least-squares minimization and root finders. Each particular method also tends to have many user-configurable parameters, and we encourage the reader to consult the scipy documentation before applying a function to their dataset.

9.2.4 Data Science

Python is an extremely popular language for "data science" work for a few reasons:

1. The ability to quickly leverage work previously done in open-source numerical Python modules, such as numpy, and the wide functionality of modules available beyond statistical modeling.
2. Excellent interoperability with low-level languages, databases, and web frameworks. This makes it easier for companies to integrate data science output in their production web environments.

3. High productivity design and reputation as a "trendy" language. Python is the most popular language for "Intro to CS" classes at a time when data science is a high growth field: supply of young developer skills coupled with employer demand. It is also generally easier to learn for subject matter experts without formal CS training, as compared to C++ or R.

Two important libraries to learn for data science-related work are `pandas` and `scikit-learn`. Pandas is a package designed to work with "relational" or "labeled" data. It provides a great amount of functionality for statistical work that users may otherwise perform in Microsoft Excel, xmgrace, or R, and builds upon numpy. It also allows users to perform operations similar to SQL queries on databases, quickly retrieving complicated sets of data. The most common object used with pandas is its `DataFrame`, that is a generalized 2D mutable object allowing for mixed types—in short, a lot like an Excel spreadsheet but with a lot more functionality available via a variety of methods provided by pandas and the Python standard library. Scikit-learn is a machine learning library that provides packages with techniques such as Bayesian regression, support vector machines, Gaussian processes, decision trees, clustering, ensemble methods (including random forests), principle component analysis, factor analysis, and basic neural networks. It is often applied to problems of classification or regression. Both pandas and scikit-learn come with many methods for importing and preparing data for analysis. There are also many advanced machine learning libraries, in particular for neural networks or deep learning, which often build upon the Python module Theano. That is an advanced module, but a starting point if you wish to pursue this topic past the tools provided in scikit-learn. For specific examples, we recommend the reader consult the current web pages for pandas and scikit-learn.

9.3 Python Extras

9.3.1 Python Style

Python best practices encourage a coding style that is documented in "PEP 8," formally known as Python Enhancement Proposal 0008. Many editors, such as Spyder (an IDE discussed in Section 9.3.5), will automatically check your Python code for PEP 8 conformance as you are writing it. This combination of a recommended style guide (as opposed to just a language standard), tools for checking it on the fly, and whitespace being part of the language standard mean that Python written by one author will often be surprisingly readable to someone else. This is one of the design principles of Python, as mentioned in PEP 20, "The Zen of Python": Readability counts. For example, PEP 8 recommends using four spaces for indentation, as opposed to two spaces or tabs. This makes your code more readable to other Python programmers—they will be used to that style—and ensures that indentation levels and types are not mixed across a large code base. Python 3, in fact, disallows mixing of tabs and spaces for indentation. Whitespace is also recommended around assignment operators; for example, use "`a = b`" instead of "`a=b`". Whitespace can elongate a line of source code, but PEP 8 also recommends limiting yourself to 80 characters per line. For line continuation, PEP 8 recommends using implicit line joining by having parentheses, brackets, or braces joining the statements across lines and then aligning the continuation with the opening delimiter. If using Spyder, the automatic PEP 8 linting will help with correctly handling your whitespace. Naming conventions generally used are short, all-lowercase names for modules, `CapitalizedWords` for classes, and `lowercase_with_optional_underscores` for most other things such as functions, method names, or variables. In fact, now is a good time to check out The Zen of Python in full. At an interactive Python prompt, execute:

```
import this
```

Finally, know that these are style *guidelines*, not language standards. PEP 8 also says "know when to be inconsistent—sometimes the style guide just doesn't apply." Python, and its style, is meant to be programmer-friendly for high productivity. One last task left to the reader is to import antigravity in Python 2.7+ using a local python interpreter. You will know when you have succeeded.

9.3.2 Parallelism

If doing serious number crunching, or interfacing Python to a parallelized C or Fortran program, you will eventually want to use all of the CPU cores at your disposal. Python has many different ways of supporting this. We will focus on two modules: `multiprocessing` in the standard library and `mpi4py`, a third-party library. Multiprocessing is appropriate for shared memory parallelism, while Message Passing Interface (MPI) programming is appropriate for both shared and distributed memory parallelism. Aside from being in the standard library and therefore instantly available to anyone using Python, `multiprocessing` is popular because it largely avoids the Python "Global Interpreter Lock" (GIL) by using process-based instead of thread-based parallelism. The GIL is a lock, or a form of permission, that must be held by a thread before the Python interpreter allows it to access a Python object. It may only be held by one thread at a time; hence, avoiding the GIL improves performance. For the following parallel Python examples, we will assume that you are executing from a script instead of entering commands one by one at an interactive prompt. This is important because the multiprocessing module should have commands executed from within a "`if__name__=='__main__'`" construct in order to guard the "entry point" of the module; essentially, if a new process were spawned, you would not want it to believe it was the original process and re-execute all the commands in your script. Windows and Linux handle spawning processes differently, and it is important to understand that behavior if you write more complex multiprocessing code. Let's start with "Hello, World."

```
import multiprocessing

def worker(n):
    my_name = multiprocessing.current_process().name
    return 'Worker %d using process %s' % (n, my_name)

def hello_world():
    pool = multiprocessing.Pool(2)
    results = pool.map(worker, range(4))
    pool.close()
    print(multiprocessing.active_children())
```

```
    for message in results:
        print(message)

if __name__=='__main__':
    hello_world()
```

The above example spawns two processes by creating a `Pool()` and creates a `map` of four workers to use those processes. The results of the worker functions are collected by making the pool `close`, after which the results are printed by the main thread. This is one approach that can be used to parallelize numerical computation. How the computer schedules work on those processes is machine dependent: for this example, a simple task, all the workers may be on the same process (e.g., using Windows). Repeated invocations will lead to different load balancing. Try it a few times yourself, then let's move on to an example using `mpi4py`. You will need to ensure that `mpi4py` is installed, along with its dependencies; typically, this is best attempted on Linux and, depending on the system at your fingertips, may be easier to try on Python 2.7.

```
from mpi4py import MPI
comm = MPI.COMM_WORLD
rank = comm.Get_rank()
size = comm.Get_size()
name = MPI.Get_processor_name()
print("Hello, World! I am process %d of %d on %s" % (rank, size, name))
MPI.Finalize()
```

Ultimately, MPI-based parallelism is more common for scientific computing as it is the *de facto* standard for communication across nodes on supercomputers. It is common to see traditional high-performance computing (HPC) applications make use of MPI bindings for C and Fortran, which are typically faster for equivalent codes and demonstrate better scalability than Python does (especially during program initialization). The above example is based on mpi4py, but Python has multiple MPI libraries and the user should explore other useful options. There are also other options for concurrency-related programming tasks in the Python standard library, such as the `threading` module.

9.3.3 Matplotlib

Matplotlib is a Python library that allows a user to make simple plots from array or list data and output them as images in only a few lines of code. It makes heavy use of numpy and provides good performance for large arrays of data. The library can easily generate simple-looking plots, histograms, vector fields, and even three-dimensional contours. Below is an example of a matplotlib histogram, using a real-world dataset in the variable `resListA0` and range variable `numResX`. This code will not execute if you paste it in the interpreter—instead, we encourage the readers to generate their own list of random numbers between 0 and 200.

```
from matplotlib import pyplot

pyplot.hist(resListX0, bins = numResX, range = [0.5,numResX+0.5], align =
'mid', facecolor = 'g', rwidth = 0.9, alpha = 0.75)
```

```
pyplot.axis([0.5, numResX+0.5, 0, 15])
pyplot.xlabel("Residue on X")
pyplot.ylabel("Frequency")
pyplot.grid(True)
pyplot.savefig("contact_map_example")
```

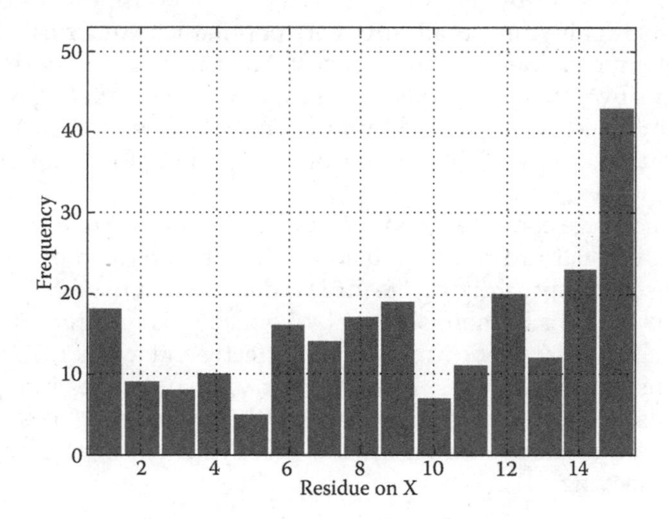

9.3.4 Porting from 2 to 3

In the design and release of Python 3, the development team made a significant decision to break backward compatibility with significant parts of Python 2. Ultimately, it was to make Python 3 a more consistent, easier-to-use, and more modern language. If just beginning with Python, it is best to program in Python 3. There are many tools created to help with porting code from 2 to 3, such as the module `future`, the futurize script, and the 2 to 3 script. If using a recent version of Python 2, you can also import some Python 3 syntax using the "`from__future__`" construction. We now review the most important "2 to 3" differences related to computational science.

The `print` function has changed. It now requires parentheses to enclose its argument, instead of accepting standalone strings. Division has changed. In Python 2, dividing two integers would result in an integer (rounded down). In Python 3, it will result in a `float`, as if the two operands were floating-point numbers. This is critically important for numerical work, especially if you are using a script someone else created in Python 2 to do work in Python 3! A syntactically incorrect print statement will raise an error, but the change in division will silently give you different results. Iterating, such as in a `for` loop, has also changed. In Python 2, a commonly used generator is `xrange`, but in Python 3, the function `range` should be used in its place (and has functionality equivalent to Python 2's `xrange`). Finally, string handling has changed, with all text strings now being Unicode by default. In Python 2, the `str` type handled text and binary data, but Python 3 has a `bytes` type to handle binary data. This change may affect users processing large amounts of strings for different purposes in their program. Python 3 has many other backward incompatible changes, but they are often easier to deal with and extensively documented. When porting from 2 to 3, an extensive test suite should be prepared and code validation tools utilized.

9.3.5 Third-Party Tools

The examples in this chapter were created using Spyder, an IDE in Anaconda, which is a Python distribution created by Continuum Analytics, and Eclipse, a general-purpose IDE. Another popular Python distribution is Canopy by Enthought, and another frequently referenced IDE is PyCharm by JetBrains. All major editors used for programming support Python syntax highlighting. Some are more popular for numerical work, and others, for web site development. These authors suggest Anaconda, as it is designed for numerical computing and by default includes a great many Python packages not available in the standard library, such as scipy and numpy. The Spyder editor in Anaconda embeds IPython, a powerful interactive shell that is becoming popular for its support of web-based notebooks.

Static analysis of source code is a good second check on your codebase, in order to find hidden errors or potential future issues before things become too problematic. In short, static analyzers assist in minimizing "technical debt" when used regularly. One of the best static analysis tools for Python is pylint by Logilab. It is useful to run pylint on your code, even if the code appears to be functioning perfectly and to be PEP 8 compliant; pylint may have some insightful messages. Also popular, and sometimes integrated into IDEs, are pyflakes and its style-conscious cousin flake8. We recommend flake8 as an intermediate option but is not quite as powerful as pylint when your editor does not have on-the-fly style and syntax checking.

Further Reading

https://www.python.org/doc/.

http://docs.scipy.org/doc/.

http://pandas.pydata.org/pandas-docs/stable/.

http://scikit-learn.org/stable/documentation.html.

http://python.net/~goodger/projects/pycon/2007/idiomatic/handout.html.

H.P. Langtangen. *Python Scripting for Computational Science*, 3rd Ed., Berlin, Heidelberg: Springer-Verlag, 2009.

10

Prototyping

Charles Lena

CONTENTS

Most industries benefit from hastened decision making through the use of additional input data. Whether in the form of consumer marketing surveys for usability or actual physical test cases falling over on a test desk, prototyping is used to more rapidly inform ourselves of needed design decisions. As engineers and scientists, the adoption of rapid prototyping as a method of informing software design and deployment should not surprise us. This chapter will lay out the reasoning as to why and when prototyping will save both time and headaches in the design process, especially within the realm of scientific software development. It will also present a few possible prototyping workflows.

10.1 What Is Prototyping?

First, we will address the what of prototyping: prototyping refers often to the consideration and refinement of software design requirements while factoring in customer interactions, graphical interfaces, and usability of an accessible public interface (API). Furthermore, prototyping now often implies "rapid" prototyping, in which prototyping is done with sprint-like development periods while in constant communication with onsite customers. In this regard, it is more than a little awkward to talk about this in regard to most scientific software, which more often than not will lack any graphical user interface (GUI) and will only have two customers—you and, by proxy, your group's advisor. Of course, the goal of most scientific software are scientific discoveries and their resulting publications. This situation, while not ideal in the world of rapid prototyping, certainly should inform the design of the scientific software.

When I say customer interactions, we specifically refer to the usability of the software platform in development. If a piece of software is difficult to use in consumer space, consumers will simply use a different piece of software—unless there is no other. In scientific space, we all try and produce novel pieces of software; there are often no alternatives that do the exact process for the first publishing cycle. This means that customer interactions are often ignored in favor of other design input. A personal example comes from my work in the field of density functional theory—more specifically, I work on the Pseudopotential Algorithm for Real-Space Electronic Calculations (PARSEC). This code has computational advantages compared to others in the same space, such as Vienna ab-initio simulation package (VASP) and Quantum Espresso—but science and scaling aside, the real challenge with PARSEC is usability. The learning curve to use any of these pieces of software on complex simulations is not simple, and it is not uncommon to shop around in the space for software that provides the most direct, documented route for the desired research simulations. Ideally, we could design our software with user interactions and usability as first class concerns.

Likewise, most scientific software focuses on the algorithms and calculations, and even on their presentations and visualizations, before any sort of graphical user interface. Lacking graphical interfaces, most programs will simply resort to set parameter files and input datasets. This is not unreasonable—GUIs are a large amount of work for functionality that can be effected with much simpler to develop command line interfaces. Nevertheless, the lack of a GUI removes some of the obvious strengths of rapid prototyping as it is applied in the most prevalent consumer software space today.

In scientific labs, we often have neither a focus on user interactions nor usability—bleeding edge scientific research offers up many new and unique approaches to scientific problems, and publishing drives the automation of these approaches inside of software platforms. This lack of architecting is common in scientific code development efforts, as the focus is not on the code itself but on the science behind the code. We gain much from prototyping through forcing development speed and clarification of requirements. Prototyping informs our design by speaking to the requirements not understood at the outset of the effort. Another benefit which should not be overlooked is that prototypes offer paths to understanding for new group members. This is true whether a group continues modifying prototypes or creates many small disposable prototypes, that choice being a contentious choice in some camps [1].

When prototyping, we must consider the end goals of the prototype. Obviously, we want to better the development process of our product, reducing both the total time of the product development and the amount of work required to reach it. In general, the act of prototyping is a positive one that builds good habits. In that regard, it is similar to unit testing as an activity—positive but often underutilized in prototyping scientific software.

10.2 Choosing the Language for the Prototype

We must also choose the language for the prototype. Rarely a given, we choose languages based on features providing ease of development for functions and data structures, which should comprise the majority of algorithmic work. Some people, especially those trying to reduce the danger of a prototype becoming a time sink, can work in the same language as the production software. This is certainly not a requirement and might actually be a hindrance if the main language project is to be in C, Assembly, or another language without much "syntactical sugar" to aid in development.

For the prototyping of most scientific software, I prefer Python, which I suggest for a few main reasons. First, it is easy to relate to Fortran, C, and C++ and can even prototype the Message Passing Interface (MPI) variants of these languages in a pinch. Second, the read-eval-print loop (REPL) and mature development environment allows for rapid development even for relative beginners—especially given that the language reads like English-from-a-six-year-old, allowing others to comprehend and follow later. Third, Python has a huge amount of functionality built in or available; this can reduce wasting time by reinventing the wheel. Finally, the scientific community already widely uses Python for product code, and many powerful, useful modules which can be used in the product code have APIs that are freely available.

My preferences aside, I encourage you to use whatever works. The point of prototyping is to inform design in a very fast manner. Fighting language restrictions in the prototype is not a useful expenditure of time—choose a language that can properly map to the product code. Almost as important, choose a language that can solve the problem in an acceptable amount of time. My Python code will almost certainly be slower than my compiler optimized C routines. If this is slow enough to prohibit execution of even example problems, we have chosen the wrong language.

10.3 Fighting Knowledge Gaps

Similarly, fighting knowledge gaps through prototyping provides dangerous distractions to the stated goals of the activity. Novel algorithms are going to be difficult. Learning a new language is difficult. Avoid fighting both an algorithm and a language at the same time. Although temptations to be clever abound, avoid any sort of cleverness in prototype code. Many reasons for this include the idea that when prototyping, we are looking to uncover pitfalls of our design paradigms—covering up flaws with cleverness happens, and hindsight of such mistakes is a hugely painful experience. This holds true for production code as well. The compiler will often try clever things as well—cleverness in the actual code is too often synonymous with unreadability. Finally, always document and comment the prototype. It might become production code—at the very least, someone (probably you) will have to read it later.

Another great tool to use in rapid development is an old combination—pencil and paper. A whiteboard works well when prototyping as a team. Diagramming program flow does not only help the development of product code; the speed of prototyping greatly increases as we have to rewrite fewer code blocks. Our quick sketching of data communication patterns and paradigms before touching a keyboard will almost always catch errant ideas before we waste any further time on them. Figure 10.1 provides a workflow example for a prototyping brainstorm session.

In an example from my own experience, there was a prototype project done in C++11. It was an evolutionary prototype for a new method of optimization and was therefore intended to turn into actual product code. Given that most of the new students and post-docs at the time worked primarily in Fortran and the preexisting codebase was in Fortran, the choice of coding in the advanced C++ standard, while intriguing to the developer, left other members of the group struggling to engage. It also limited the number of clusters on which the software could compile and run. As the group used several different computational clusters, this limitation put serious strain on the group. Soon thereafter, the next stage of the prototype was done in C++14—the hole deepened. When the originator graduated, they left a legacy of code that was most likely mothballed.

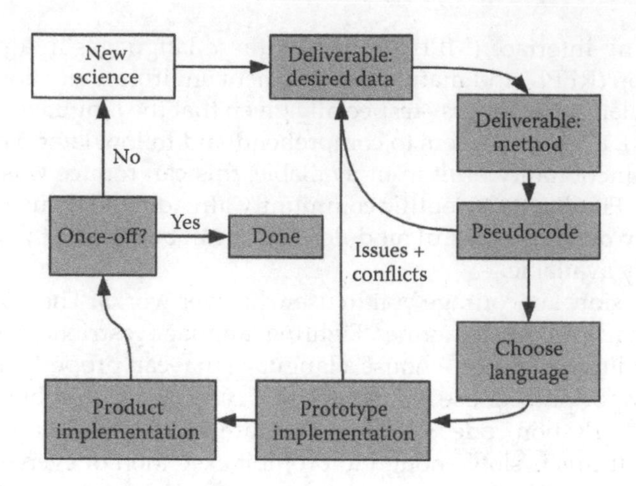

FIGURE 10.1
The workflow depicts prototyping a full product from some theoretical scientific start point. The desired data create a deliverable, which informs almost every other decision. Often, the simplest solution will differ between deliverables—identifying problem points quickly is one of the strengths of prototyping.

10.4 When to Prototype?

Maybe even more important than *what* tools to use is the *when* of prototyping. Unfortunately, social research still cannot definitively answer the question of when prototyping provides the most benefit, but in a comprehensive study, most groups reported positive reactions to results when prototyping [1]. In more direct terms, people who prototype like what happens when they do. They tend to feel like their efforts are effective and their coding is more efficient.

There are very few situations that are not appropriate for quick code sketches, which are the most basic prototyping forms. In many simple cases, a prototype will likely become the product code. In more expansive cases, the prototype allows us to answer questions and test relationships between functions, data structures, and new features before committing to product code. There are still many valid concerns that may cause your supervisor grief—often, the observation that the prototype sometimes takes more time than the finished product bothers people unfamiliar with the process. The reality is that the finished product, being subject to large rewrites and hold-ups, thanks to unfinished design, would have taken much longer without the prototype. Hopefully, the resulting product speaks for itself.

10.5 Software Design

Prototyping effectively involves multiple mannerisms that center around software design. We could write an entire separate book on software design, so this is by no means an exhaustive treatment. Treat the following more as guidelines rather than rules. These guidelines should help direct your efforts in more effective mannerisms.

Treat early prototyping like a brainstorming session. We are trying to lay out the relationships and rules of our product. See, for example, another workflow for prototyping in

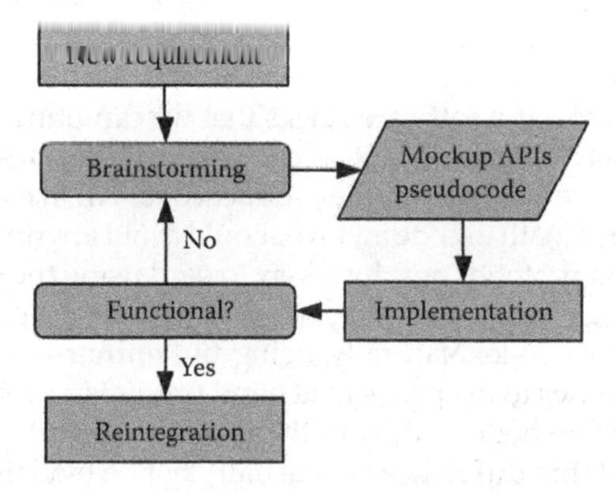

FIGURE 10.2
This workflow depicts prototyping a new feature for inclusion in some existing product. Straightforward, there is real talent in recognizing poor decisions before wasting large numbers of man-hours on their implementation. Prototyping seeks to highlight those missteps before they happen.

Figure 10.2. Prototyping takes the brainstorming and turns it into quick functional code—doing them at the same time is the closest we can reach to a customer–developer relationship that is used when prototyping GUIs in consumer space. We must not forget that we, as researchers, are the primary customers of our own code—having honest discussions about our own needs is a requirement!

Required functionality informs required program units. Knowing what the software needs to do dictates what needs to get done. Separating functionality into smaller pieces means shorter routines, allowing for much faster development in two ways. Shorter routines—comprising more modular code—are the standard in most code shops. Larger routines require more reworking and are harder to comprehend, adding potential for unnecessary slowdowns. It is psychologically easier to discard a short routine than a large routine when the need arises. With shorter routines, we maintain focus on our own deliverables.

Well-organized prototypes lead to well-designed product code. One of the hardest activities when rapidly iterating is the locking of interfaces. On the one hand, by understanding how program units relate to each other, we can infer the required behavior and inform design of interfaces. This understanding is extremely important—poorly thought-out interfaces are a surefire way to waste time rewriting after redesigning otherwise simple code blocks. On the other hand, simple straightforward interfaces that have a clear purpose and relation to other interfaces will avoid confusion about the purpose. It leads to not only cleaner code but also cleaner programs. Cleaner programs, whether at the prototype or product stages of coding, always serve a noble purpose as examples to those developers who come after.

With a goal of cleaner programs, it should be evident that the avoidance of unnecessary features is key to time-saving. The importance of removing erroneous features during prototyping bears mention as well—I am hard-pressed to think of a time when I sat down with any of my peers to develop anything larger than the smallest of additions without adding "cruft" that should have been avoided. Recognizing this cruft, and removing it in the prototype phase, along with proper documentation, will provide a barrier to people repeating the mistake in functionality further down the line. Again, saving time and bettering the product should remain the goal. The danger in this lies in tendencies for some to almost obsess over basic banalities like formatting of text—this is not *carte blanche* to waste time on formatting. Save that for the end.

We should always work toward the product's deliverables—APIs and functionality that we can control are paramount to the end goal of product code. In this regard, we must always be conscious of the toolchains and software stacks that we can utilize. If we design a product around specific functionality that we cannot access, we are implicitly committing ourselves to larger amounts of effort when working on the product code. We like to avoid commitments—prototyping provides a quick path to code and we should avoid anything that could hinder that advance. That being said, in prototyping, always err toward using the existing framework first.

There is obviously a danger when shopping around for existing libraries to incorporate into both prototype and product code. Naturally, being unfamiliar with existing code projects that provide various amenities to the problem at hand presents a problematic decision—how to grade them. One might be better suited to the prototype, while the other may be better suited to the product, and this difference will usually not be evident until we have already made the decision. The best way to avoid this conflict of interest is to double down on the aforementioned modularization. Abstracting away the choice of certain libraries' interfaces provides a quicker way to both change to a different library and keep the actual point of the code clearer. A very clear example exists in the form of much of the Linear Algebra Package (LAPACK) and Basic Linear Algebra Subprograms (BLAS) interfaces—we often have various types of double-precision general matrix–matrix multiply (DGEMM) or double-precision symmetric matrix eigenvalue solver (DSYEV) calls that we hide inside other functions akin to wrappers. Later, when there are further optimized calls that we want to use, changing what the guts point to is far easier than actually having to change the bit of the routine they exist in. Once again, like with formatting and textual organization, it is very easy to add too many layers of abstraction and create a monstrous task where one never really existed. Unneeded cleverness can also manifest as a tendency to overly abstract a problem.

At the most extreme, using too much abstraction results in using too much generalization. This problem is obviously a double-edged sword—programmers tend to strive for generalization, and that tendency can get goal-oriented code projects in trouble. There are a lot of one-off code projects for the purpose of publications—this is not inherently bad, but as part of design, we should utilize generalization when possible to reduce the amount of recoding we must do. To clarify, generalizing everything all the time is most definitively a waste of time. Generalize to the level that a project can be useful. Likewise, generalize a prototype to the level that you intend the project to reach. Finding the proper amount of generalization in a project will allow the developer to produce a large amount of science from a relatively small amount of effort expended on code, and well-designed prototypes generally provide code that other groups will want to use.

10.6 Pitfalls and Traps

As with any other approach, pitfalls and traps exist. Choosing the language for the wrong reasons is a common pitfall, and is common in many settings, graduate labs included. Wanting to learn a new language feature and using the prototype as an excuse to cultivate a new skill are also not strong reasons to choose a language at the cost of other factors. This error can compound quickly as struggling with the language will hurt over the course of the project, influencing the amount of rewrites and restructuring required. Worse yet, new features in prototype code commit any other developers to develop their own comprehension of those new features. This can be unreasonably hard for beginner developers.

Another aforementioned trap involves being overly clever in the prototyping language. A simple way to describe an effective prototype would be "working pseudocode." It should convey the message and proceed to provide implementations that strive to offer straightforward translations to the product language. An example of being overly clever is using unique mappings when iterating instead of simple ranges. A specific python example is using multiple nested comprehensions in place of simple loops.

Again, we look to cleanly map into whatever the product language is going to be. Having a diverse language for the product like C++11 allows for a much more extensive use of Python containers and algorithms. If the project is small enough, it may not matter—but hindsight will be 20/20 and rather painful in some of those cases where the project does not clear the "enough" benchmark.

Another case where hindsight is often more illuminating than prior planning are unit tests in prototyping. I encourage the reader to carefully digest the chapter on Testing—and to understand that the same arguments that apply to unit testing the product code apply to the prototype. We want to avoid spending the time required for 100% coverage unit tests. One hundred percent coverage costs too much development time to justify—with the exception being in evolutionary prototypes. Being able to inherit all the previous work makes the unit test coverage an easier pill to swallow. Sometimes, the best consideration is how complex the unit test will be—unit tests that are complex warrant inclusion in the prototype. Utilizing integration or systems tests, instead of comprehensive tests, will allow for the most coverage with the smallest amount of effort, especially given how quickly prototypes change. Similarly, this avenue for feature creep has a dangerous ability to become a serious time sink that does not provide a value add.

All of these potential time sinks aside, it is OK to spend more time prototyping than on writing the production code. In fact, as we have said many times, ideal prototyping reduces the total amount of time for the product code development. In the cases of very thorough prototyping, the production code will be little more than a translation and tidying up. This is most likely not the most efficient method of prototyping—if only because spending large amounts of time arguing with your bosses over the efficacy of prototyping will not generate any goodwill. Therefore, if you have extremely resistant supervisors, academic or corporate, it is better to limit the amount of time prototyping. It may be best to generate a prototype that covers the most complex parts of the routines and their interfaces—locking down these relationships is key to any successful prototype. Just like with unit testing, it is often more important to get most of it covered with speed than all of it covered in resistance. Time spent does not equate to quality. Make sure that the time you put into the prototype and transition is quality time and effort. Finally, while we provide two example prototyping workflows, for new features and entirely new applications, I encourage you to test your own prototyping flows and to adapt to the situation at hand.

Acknowledgments

I acknowledge the support provided by Scientific Discovery through the Advanced Computing (SciDAC) program funded by the U.S. Department of Energy, Office of Science, Advanced Scientific Computing Research and Basic Energy Sciences under Award Number DESC0008877.

Reference

1. V. Scott Gordon, and J.M. Bieman. 1995. Rapid Prototyping: Lessons Learned. *IEEE Softw.* 12, 1, 85–95. doi:10.1109/52.363162. http://dx.doi.org/10.1109/52.363162.

11

Introduction to High-Performance Computing Systems

Todd Evans

CONTENTS

High-performance computing (HPC) systems, harness tens to thousands of computers working in parallel in order to perform more computation in less time than would be feasible using a single computer. The computational power provided by HPC systems may be applied to problems that are too large to run on a single computer or used to reduce the runtime required to complete a problem of fixed size. HPC systems are used in a broad range of research pursuits including both scientific and commercially oriented fields. The community of people that run applications on an HPC system are commonly referred to as the *users* of that system.

The computers that constitute an HPC cluster communicate with each other over a network. The network may be used to share data and coordinate computing tasks. Each computer connected to the network of a cluster is referred to as a node. In practice, the nodes in an HPC cluster are relatively uniform in architecture and capabilities. There may, however, be some nodes with additional hardware such as specialized chips capable of performing specific computations efficiently or additional resources such as larger memory or numbers of processors.

An important aspect to the performance of an application on an HPC system is how efficiently the work it performs can be done by independent tasks that may communicate and/or share data. The ability to apply multiple tasks to a computation is known as *computational parallelism*. Parallelism is implemented by the programmer at the software level but originates from parallelism in the hardware. This chapter will first cover where parallelism arises in the hardware of HPC systems, followed by how individual applications interact with the hardware-based parallelism through software and software libraries such as

Message Passing Interface (MPI) and OpenMP, and finally discuss how an HPC system's users work in parallel through the use of a batch system.

11.1 Hardware-Based Parallelism in HPC Systems

An HPC system is organized into a multilevel hierarchy with a new mode of computational parallelism introduced at each level. This hierarchy can be defined in any number of ways, but here, we decompose it into four levels:

1. Node-level parallelism—An HPC system has multiple nodes working in parallel.
2. Processor-level parallelism—A node has multiple processors working in parallel.
3. Core-level parallelism—A processor has multiple cores working in parallel.
4. Instruction-level parallelism—A core can execute multiple instructions in parallel.

A schematic of this hierarchy along with the magnitude of parallelism introduced at each level is displayed in Figure 11.1. At the node level, an HPC system consists of multiple nodes connected to each other over a network. Each one of these nodes is a complete computer running its own operating system (OS). The processor level arises from the multiple sockets contained on each node, where each socket has its own processor and memory in addition to other specialized computing and I/O hardware. At the core level, each processor has multiple interconnected cores with private and shared memory capable of independent computation. Finally, at the lowest level, each core is composed of multiple subunits capable of executing instructions in parallel. The rest of this section will describe each level of parallelism in greater detail.

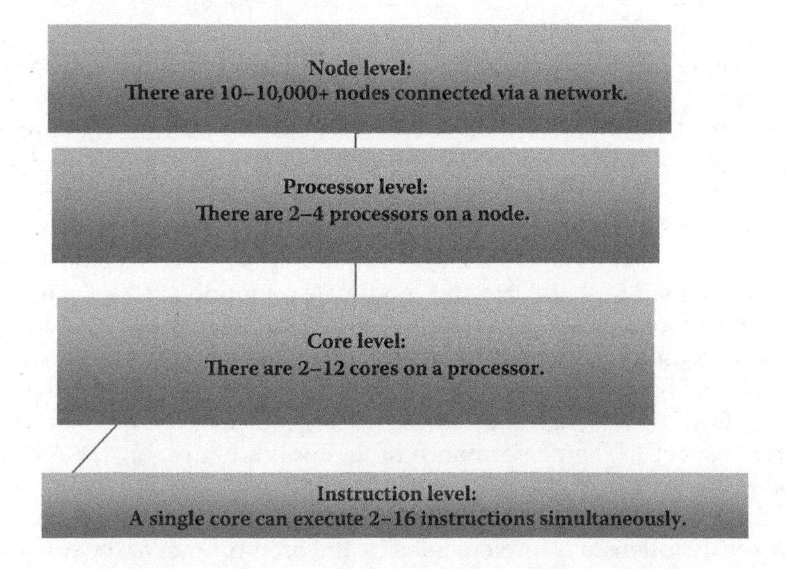

FIGURE 11.1
Parallelism appears at multiple levels in HPC systems. The lowest level is instruction execution within a core, the next level is due to multiple cores, the next is due to multiple sockets within a node, and the highest is due to nodes connected through a network.

11.1.1 Instruction-Level Parallelism

At the lowest level of the computational hierarchy are the *execution units* in the cores with which instructions are executed. Different types of processor instructions (e.g., move, add, and multiply) require different execution units and numbers of processor cycles to complete. Most modern day processors used in HPC systems are referred to as superscalar, which effectively means they can *pipeline* instructions through the execution units. Pipelining is the ability of a processor to start a new instruction before a previously issued instruction has completed. Instruction pipelining allows processors to complete multiple instructions in the same number of cycles required for a single instruction. This is what is known as *instruction-level parallelism* and represents the lowest stage of the HPC computational hierarchy.

In addition to instruction-level parallelism arising through pipelining, many processors have one or more vector processing units (VPUs). VPUs are a particular execution unit capable of executing a single instruction on multiple data elements simultaneously. The number of elements depends on the processor and currently varies from two to eight double-precision floating-point numbers (real numbers represented by 64 bits). VPUs are often utilized in programming patterns dominated by loops performing floating-point operations such as addition, subtraction, and multiplication. For example, in such a loop, a single vector addition instruction might perform the simultaneous addition of eight floating-point values, reducing the number of loop iterations by eight and, consequently, the overall runtime required for the loop. The most recent generation of processors has VPUs capable of performing multiplication and addition on the multiple elements simultaneously, which in some calculations doubles the number of floating-point operations per second (FLOPS) possible.

11.1.2 Core-Level Parallelism

The next level of parallelism in the hierarchy is introduced by the use of multiple cores on a given processor. Most modern day processors are *multicore processors* with between 2 and 12 cores, where each core has identical capabilities and can perform independent computations. In modern processors, these cores typically have their own private L1 (Level 1) and L2 (Level 2) cache and a shared L3 (Level 3) cache and main memory. There are a variety of protocols that processors use to maintain coherency between the private and shared caches and main memory.

These coherence protocols are implemented in the processor's hardware and are transparent to the programmer. The purpose of a cache coherency protocol is to ensure that the data in each core's private cache are consistent with the shared cache and memory. For example, if one core modifies the value of a particular variable, all other cores must be notified of this modification and wait to read this variable until the new value is available. The new value may have to be written to shared cache or memory or may be forwarded directly to the core's private cache that requires it. A detailed discussion of cache coherence protocols is beyond the scope of this chapter.

11.1.3 Processor-Level Parallelism

The next level in the hierarchy arises at the socket level. Commodity servers often have multiple, for example, two to four sockets within a node, where each socket has a single processor with its own main memory. A possible layout of such a node with two sockets is shown in Figure 11.2. The memory in each socket is visible to every other socket and appears to the programmer as a single memory space. The sockets are connected by on-node interconnects that allow the multiple sockets to maintain cache and memory coherence. There is, however, additional latency associated with accessing memory that is not local to a processor's socket. The access of data nonlocal to a processor is known as nonuniform memory access (NUMA). In contrast, uniform memory access (UMA) refers to accessing memory local to a socket; for example, all cores within a processor access the local socket's memory in an identical manner. The coherence implementation of NUMA and UMA processors is transparent to the programmer.

In addition, a node may have what is called an accelerator. An accelerator may be Intel's Xeon Phi Coprocessor or a graphics processing unit (GPU) such as is available from Nvidia. Accelerators have tens to thousands of cores capable of performing large numbers of floating point or integer operations in parallel. These cores are specialized for specific and very regular computational tasks, allowing them to be less complex, cheaper, and more energy efficient than a processor's cores. Conversely, accelerators are less flexible than a processor and perform poorly at general tasks such as control flow and memory management.

There is also often hardware on the node that aids in off-node communications. For example, Infiniband Architectures (to be discussed in the next section) have a host channel adaptor (HCA) or target channel adaptor (TCA) card on each node that manages most tasks related to internode communication. The HCA and TCA cards, in addition to being highly specialized for communication tasks, free up the processors during communication and allows computational and communication tasks to be overlapped.

Currently, most accelerators and HCAs are accessible by the processors through a Peripheral Component Interconnect Express (PCIe) bus. PCIe is an interconnect or bus used in nodes to interface to many different devices, ranging from hard drives to main memory to the abovementioned accelerators. Similarly, there are specialized interconnects available that connect the processors and memories of individual sockets, such as Intel's Quick Path Interconnect (QPI) and AMD's HyperTransport interconnect, and are used for tasks such as maintaining across-socket cache coherency.

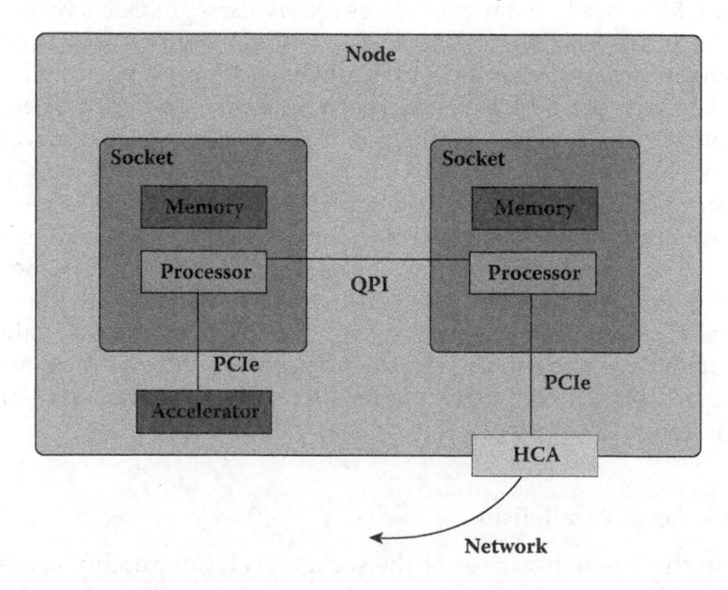

FIGURE 11.2
Possible layout of compute node in an HPC system.

11.1.4 Node-Level Parallelism

The node level is the highest mode of parallelism in the HPC system hierarchy. The nodes are connected by a network, with each node running an instance of an operating system and capable of acting as its own independent computer. The network can be organized into different *topologies*: a particular arrangement of nodes and connections. The topologies of networks with more than a few nodes also often include one or more layers of switches to connect the nodes, where switches are devices that route signals from one node or switch to another. The switches are connected so as to minimize the number of hops or distance that a signal or data transmission is required to make.

The most common topology used for HPC systems today is the Clos or Fat-Tree topology. This topology consists of one or more core switches, which only connect to other switches, and multiple leaf switches, which connect to multiple nodes and switches. The number of cables used between the nodes and leaf switches and leaf switches and cores switches is chosen so as to maintain a constant or near-constant bandwidth at all levels of the tree. For this reason, it is also often called a constant bisection bandwidth topology. A two-level Fat-tree network is shown in Figure 11.3.

Most networks are based on either Ethernet and/or Infiniband architectures. Ethernet is a TCP/IP-based networking technology used in a diverse range of contexts from connecting single laptops to the Internet to interconnecting large HPC systems. The Infiniband Architecture (IBA) is a specification for specialized hardware and software designed for low-latency communications. While Ethernet is reliant upon the OS and, ultimately, the processor to manage communications, IBA has hardware that implements multiple layers of communication along with supporting features such as remote direct memory access (RDMA) and multicasting. RDMA allows one node to read or write to a target node's memory directly without the target node's participation. This is a powerful feature as it allows one-sided communication, consequently reducing latency and creating opportunities for computation and communication overlap on the target node. Multicasting refers to communications that involve multiple nodes such as a function that broadcasts a particular variable to all nodes. The hardware implementation of multicasting has lower latency than the software implementation does.

The hardware support provided by IBA allows it to generally achieve lower latencies than Ethernet is currently capable of reaching. There is hardware support for some of Ethernet's communication stack. In particular, RDMA can be supported by hardware implementing the internet Wide Area RDMA (iWARP) protocol. In any case, Ethernet communications

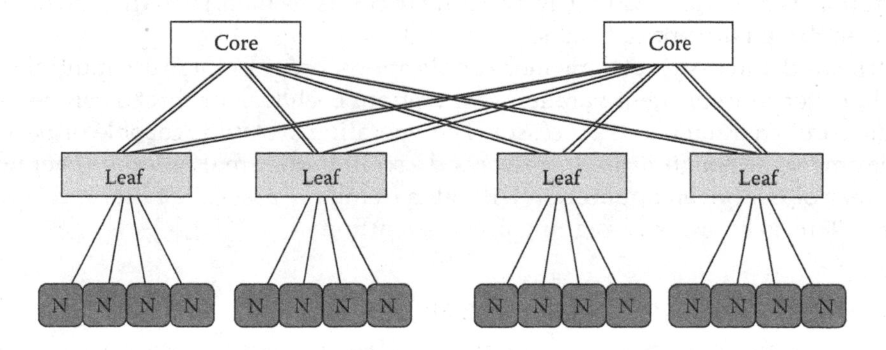

FIGURE 11.3
A Fat-Tree network topology. Both levels, node (N) to the leaf switch (Leaf) and the leaf switch to the core switch (Core), have the same number of connections and, hence, the same bandwidth available.

currently experience a factor of five to seven times the latency of IBA (1 microsecond versus 7 microseconds). Both Ethernet and Infiniband are, however, capable of sustaining comparable bandwidths, with bandwidths in HPC clusters commonly attaining 50 Gb/second.

11.2 Software-Based Parallelism in HPC Applications

A processor with multiple cores is capable of executing multiple programs simultaneously. This is what is known as symmetric multiprocessing (SMP). The cores share a memory bus that connects them to the same memory, allowing the memory to be equally accessible to all of the cores. The memory that the cores access is maintained coherently by a hardware implementation of one of a variety of cache and memory coherence protocols.

11.2.1 Memory, Processes, and Threads

Concurrently running processes may or may not communicate directly with each other. If they do communicate with each other, they may or may not have access to the same regions of memory. Regions of memory to which multiple processes can read and write are known as *shared memory,* while regions that only a single process can read and write to are known as *private memory.* Memory that is local to a node may be shared or private. Memory that is nonlocal to a node is almost always accessed as private memory and is called *distributed memory.* The vast majority of clusters lack any hardware mechanisms to maintain coherence of memory between processes on different nodes, and this must be implemented in some way by the software.

An instance of a program executing on a Linux operating system is called a process or task. In Linux, each process has its own address space in memory that is visible to only it—it is private memory. Any signal or data that are exchanged with other tasks must be explicitly sent and/or received by each task participating in the communication. Within each task, there may, however, be multiple lightweight units of execution called threads. The threads are created and destroyed within a task and can access the same address space, that is, use shared memory. Therefore, explicit communication between threads within a process is not necessary. Mechanisms must be used to avoid multiple threads accessing data asynchronously in order to avoid *race conditions.* A race condition occurs when multiple threads are reading and writing to the same data, such that the data visible to any given thread are unpredictable.

In both shared and distributed memory applications, programmers use multiple tasks or threads in order to implement parallel computation. Each task or thread can be assigned to specific cores in a multicore processor. The operating system is capable of performing this assignment, although it doesn't always do so in an optimal manner. Therefore, the programmer or user of an application will often manually associate tasks and/or threads with cores. This is known as assigning processor affinity.

11.2.2 Distributed Memory Programming and MPI

The majority of distributed memory applications developed for HPC systems use implementations of the Message Passing Interface (MPI) Standard, developed by the MPI forum (http://www.mpi-forum.org). MPI implementations provide a library of routines that can be used to coordinate tasks and communicate data between them. MPI is available for most distributed memory architectures and is language independent, although in practice, implementations exist in C, C++, and Fortran, with interfaces to these implementations developed for other languages. Multiple open-source (Open MPI, MPICH, and MVAPICH2) and vendor (Intel MPI and Cray MPT) implementations of the MPI Standard exist.

MPI does not perform the actual sending and receiving of data but rather provides an accessible public interface (API) that calls routines in the underlying network protocol, for example, TCP/IP or Infiniband, which handles the transport and reliability of data. Infiniband-supported MPI has two modes of communication: channel semantics and memory semantics. Channel semantics require that the sending and receiving tasks post a send and receive request, respectively. The send request will not complete until the receive request is posted. This allows the receiver to allocate memory for a buffer only when expecting to receive a message. Conversely, memory semantics require a dedicated buffer to be in place at all times. Memory semantics use one-sided RDMA to send and receive. Involvement of the target host's processor is unnecessary—the target has preallocated buffers ready for reading and writing by other hosts. Memory semantics require more memory but incur lower latency than channel semantics do.

11.2.3 Shared Memory Programming and OpenMP

Similarly to how the MPI specification supports distributed memory programming, *shared* memory programming is supported by the OpenMP specification. The OpenMP specification is maintained by a large number of hardware and software vendors. OpenMP implementations exist for C, C++, and Fortran, with wrappers for additional languages. OpenMP's API supports the creation and management of threads executing in parallel.

Data in parallel regions of OpenMP-supported code can be declared *shared* or *private*. Shared data are visible and accessible to all threads, while private data are a copy of a variable that is visible and accessible to each thread. Synchronization is often required in both cases. Shared data should be accessed in a way that is consistent between all orderings of the instructions executed by the threads because execution is asynchronous. The private data must somehow be communicated to the other threads at the beginning or ending of the parallel region of code.

OpenMP also supports three different types of *scheduling* in parallel regions of code involving loops. Static scheduling assigns an equal number of loop iterations, or chunks, to each thread in a round-robin fashion. Dynamic scheduling assigns a chunk to each thread, with each thread returning for new chunks after it completes its assigned iterations. Guided scheduling is similar to dynamic scheduling; however, the size of the chunks decreases exponentially each time a thread returns for more work. Static scheduling has very little overhead and should be used for tasks where each iteration requires a similar amount of work. Dynamic and guided scheduling should be used where some iterations take significantly longer than others.

11.3 Batch System-Based Parallelism in HPC Systems

Batch systems provide a mechanism for users of HPC clusters to concurrently submit computational work requests and resource specifications (number of nodes, runtime, etc.)

necessary to complete the requests. These work requests are referred to as *jobs*. A user is able to submit a single or multiple jobs to the batch system, which will subsequently manage the job startup, execution, and completion without user intervention. The batch system alleviates the need for users to actively wait for resources to become available and enables the jobs to run in parallel.

11.3.1 Job Schedulers

A job scheduler on an HPC system manages the execution of jobs, where a job is composed of one or more applications to be executed. The scheduler queues jobs submitted by the users and determines their order of execution while accounting for allocatable resources. There are many policies and associated factors that can contribute to the order of job execution, and every scheduler is slightly different. Just a small sampling of available job schedulers is listed below:

- Moab Cluster Suite with TORQUE
- Portable Batch System (PBS)
- LoadLeveler
- Condor
- Simple Linux Utility for Resource Management (Slurm)
- Univa Grid Engine (previously known as Sun Grid Engine)

For purposes of this section, when implementation-specific details are mentioned, the Slurm scheduler nomenclature will be used. The Slurm methods and examples described here have direct analogs in other job schedulers. There are two categories of scheduling types available in Slurm that encompass most capabilities of other schedulers: basic and multifactor. Basic is simply a first in, first out (FIFO) scheme, where the first job submitted to the queue is the first to begin. Multifactor has many more options to determine in what order, or with what priority, a job will be run.

11.3.2 Job Priority

A job's priority determines when it will be run by the scheduler, with higher priorities run before lower priorities. In Slurm (and similarly in other schedulers), there are five major factors used to calculate a job priority, all of which range from 0 to 1.0:

1. Age—This factor increases based on how long the job has been in the queue and in a state ready to be run.
2. Fair-share—What percentage of a user's or project's share of the total resources has been consumed in a given time window is a factor in priority. There are many ways to calculate this and determine the time window, but typically, underserviced users and projects are granted higher priority than overserviced users and projects.
3. Job size—The number of nodes a job requests. This may be configured to give higher priority to jobs requesting greater or lesser numbers of nodes.
4. Partition—The partition a job is submitted to may modify the job's priority. A partition is a set of nodes defined in the scheduler configuration and is typically

established based on resource differences (large memory, presence of accelerators) or by ownership of the resources. There can be multiple partitions on a HPC system.

5. Quality of service (QOS)—Special classes of groups, users, and jobs can have their priorities and resource limits modified based on QOS values. Jobs that use a particular quality of service may have a higher or lower priority.

Slurm computes priorities from a linear combination of the above five factors:

$$\text{Priority} = w_A\text{Age} + w_F\text{Fairshare} + w_S\text{Size} + w_P\text{Partition} + w_Q\text{QOS},$$

where w_x is a weight assigned to each factor. All of the above factors are set in the configuration file in Slurm, `slurm.conf`.

11.3.3 Additional Scheduling Methods

Running jobs strictly in the order of their priority has limitations that can be overcome by employing additional scheduling techniques. With execution times based strictly on priority, the jobs are run in the order of their priority as soon as the requested resources become available. This may leave many nodes idle on a cluster. For example, assume that a 100-node system has Job A running on 51 nodes for one day. Job B requires 50 nodes and is the next highest priority job in the queue. Forty-nine nodes will be idle for the entire day because all lesser priority nodes must wait for Job B to begin before they are considered for execution. *Backfill scheduling* solves this problem by searching for and executing jobs with lower priority that can run on the available resources and complete before the highest priority jobs would be able to start. The scheduler can compute this because it knows the start time and requested runtime of all jobs in the queue.

Some combination of jobs may consume all of the resources and have runtimes of hours or more while some jobs submitted to and waiting in the queue may run only seconds or less. It may be important for a given HPC cluster's workload goals that those jobs with very short runtimes be allowed to run much more frequently than waiting for the long-runtime jobs to complete would allow. *Gang scheduling* circumvents this problem by scheduling time windows in which different jobs are allowed to run on the same resource but in different time windows. In a system with gang scheduling enabled, a given job will run for a preconfigured time window and then the scheduler will check if there are any waiting jobs that can run on those resources. The scheduler will suspend the currently running job and begin executing the next job for the configured time window. There may be numerous jobs subjected to this scheduling and sharing compute resources.

11.3.4 Job Submission Scripts

A job, or batch, submission script is used to convey to the job scheduler all of the information it needs to schedule, execute, and record accounting information for a job. A job submission script may require none or all of the following to be specified in the script when submitting a job, depending on what defaults are set for a given cluster:

1. Queue—The queue or partition that a job is submitted to. A queue will correspond to a subset of nodes available on a cluster, and a job submitted to it may be subjected to restrictions such as minimum/maximum node counts and runtimes.
2. Nodes—The number of nodes to allocate for a job.

3. Cores/tasks—The total number of cores or MPI tasks requested for a job. The exact definition of this parameter is dependent on the scheduler and cluster configuration. It may define the number of cores per node to use for a job or it may define the total number of cores to use. The layout of those total cores across the requested nodes will depend on the cluster and scheduler configuration.
4. Runtime—The maximum time the job will run. If the job runtime exceeds this, the job will be terminated by the scheduler.
5. Output/error files—stdout and stderr from the job will be directed to these files. This enables users to track the progress of their job.
6. Dependencies—Delays the start of a job until previously specified jobs have completed successfully.
7. Allocation/project—This specifies the account to which a job is charged. This information is used both to compute the fair-share factor in the job scheduler and to limit the compute time granted to a project. Once this limit is exceeded, users will no longer be able to submit jobs to the project.
8. Command line(s)—These are the actual commands to run during the job. They are typically executed on only a single node allocated to the job. It is then the task of the MPI libraries or other scripts to distribute the commands across allocated nodes.

Items 1–7 in the above list are each specified by a single line in the submission script. Item 8 could require any number of lines and use any valid script commands. An example of a Slurm job submissions script is shown in Figure 11.4.

An example of the output from the `ibrun hostname` command in Figure 11.4 follows,

```
c445-003
c445-004
```

where the command was run once on each host (once for each MPI task).

Most clusters will provide a script, such as `ibrun` in Figure 11.4, to aid in the distribution of the commands and MPI tasks over multiple nodes and cores. In this example, `ibrun` ran a single MPI task on each node, where each task called the `hostname` command. Note the flexibility in the types of commands that can be run. A command may be a single command run by each MPI task or a single application using MPI to distribute and coordinate

```
#!/bin/bash
#SBATCH -J myjob          # job name
#SBATCH -o myjob.o%j      # output file name (%j expands to jobID)
#SBATCH -e myjob.e%j      # error  file name (%j expands to jobID)
#SBATCH -N 2              # total number of nodes requested
#SBATCH -n 2              # total number of mpi tasks requested
#SBATCH -p development # queue (partition)
#SBATCH -t 01:30:00       # run time (hh:mm:ss) - 1.5 hours

ibrun hostname            # run the hostname command in each MPI task
ibrun ./a.out             # run the MPI executable named a.out
```

FIGURE 11.4
Example of a job submission script.

its work across tasks. There might even be many different commands and applications, each of which is running on different numbers of nodes and cores. It is entirely dependent on a job's required workflow.

A common workflow is to run a serial application (requiring a single core) across numerous nodes and cores allocated to a job. Each of the commands issued is identical but feeds different inputs to the application and hence may generate different outputs. Such a workflow may be used in a sweep of a parameter space, for example. This workflow is considered *embarrassingly parallel*; there is no communication required between the tasks (serial commands) that are executing simultaneously. If an embarrassingly parallel job takes 10 weeks on a single core, it will take approximately 1 week on 10 processors.

12

Introduction to Parallel Programming with MPI

Jerome Vienne

CONTENTS

The Message Passing Interface (MPI) has been the *de facto* parallel programming model for the past two decades, and has been very widely adopted to write high-performance scientific applications. What can a programmer do with MPI? A programmer can call functions from the MPI library to organize parallel computations on multiple processors and coordinate the communication between those processes.

This chapter will provide you with the basic foundation for understanding and building MPI codes. Before studying how to work MPI, I believe it is important to learn about its origin. Section 12.1 provides a brief historic introduction for understanding why and when MPI was created. Section 12.2 introduces you to the first MPI commands, before explaining how to build and execute an MPI code. But as MPI uses messages to exchange data between MPI processes, it is important to know how to do it. This is what we will start to see in Section 12.3, which will introduce point-to-point communications before describing collective communications in Section 12.4. As learning is always better with examples, an example of code using MPI to compute π is provided in Section 12.5. If you are interested in learning more about MPI, Section 12.6 provides a list of books and websites that can help to improve your knowledge.

12.1 The Origin of MPI

During the 1980s, many companies were already building and selling parallel computers. Their programming environments were composed of a sequential programming language (C or Fortran) and a message-passing library to allow the different processes to communicate. Unfortunately, each builder had their own message-passing library, which led to many issues. In fact, a parallel code developed on an nCUBE/10 could not compile and run on an Intel iPSC. Rewriting/modifying the code was required every time a different architecture was used. It was a big concern because parallel codes were not portable. There was a need to create a standard for all these different message-passing libraries.

This is what happened with the creation of the Message Passing Interface (MPI) standard and its initial version in 1992. This standard defined an accessible public interface (API) that is now used by all MPI libraries to exchange messages. Version 1.0 of the standard appeared in 1994. The MPI standard is still improving, and multiple versions have been released over the years. The most recent version is 3.1, released in June 2015.

12.2 A Little MPI Program to Start

MPI is a library that uses specific instructions but also specific commands to compile and execute the parallel code. In this section, we will introduce and explain the first instructions that you need know, but we will also learn how to build and launch an MPI code.

12.2.1 From Serial to Parallel

First, let's take the classical "Hello world" example, which can be found in the *C Programming Language* book by Kernighan and Ritchie.

LISTING 12.1 "HELLO WORLD" IN C

```
#include <stdio.h>

main ()
{
     printf ( " Hello world \n" );
}
```

If you compile and run this program, only one process will print "Hello world."

The usage of MPI will allow you to execute your program on multiple processes at the same time. Thus, a parallel version of the classical "Hello world" can be written as shown in Listing 12.2.

LISTING 12.2 MPI VERSION OF "HELLO WORLD" IN C

```
#include <stdio.h>
#include "mpi.h"

int main (int argc, char *argv[]){
    int rank, size;
    /* starts MPI */
    MPI_Init (&argc, &argv);
    /* get current process rank */
    MPI_Comm_rank (MPI_COMM_WORLD, &rank);
    /* get number of processes */
    MPI_Comm_size (MPI_COMM_WORLD, &size);
    /* print the rank and number of processes */
    printf("Hello world from process %d of %d\n", rank, size );
    /* ends MPI */
    MPI_Finalize();
    return 0;
}
```

This version contains a lot of new functions. Don't worry, we will look at them in detail. But before talking about these functions, two things need to be highlighted:

- First, a new header file called mpi.h was added. This header contains not only prototypes of all MPI functions but also all macro and type definitions required to compile an MPI program.
- Second, all the MPI functions start with the string "MPI_". In fact all functions, macro, and type defined in the MPI start with this string. Now that we are finished with observations, let's look at the MPI functions used inside our code.

12.2.2 MPI_Init and MPI_Finalize

MPI_Init is the first function to initialize the MPI environment.

```
int MPI_Init( int *argc, char ***argv )
```

The arguments *argc* and *argv* are pointers to the argument to the main function. For MPI implementations compliant with MPI-2 or newer standards, MPI_Init also accepts NULL as input parameters. One important thing to know is that the communicator *MPI_COMM_WORLD* is created during this call.

In fact, all MPI ranks or MPI tasks need to be identified. For this, the MPI library uses a nonnegative integer and attributes a unique number, starting at 0, to each process involved. All these processes are part of a group called a communicator, and the name of this initial communicator is called *MPI_COMM_WORLD*. Each MPI process can be mapped on one

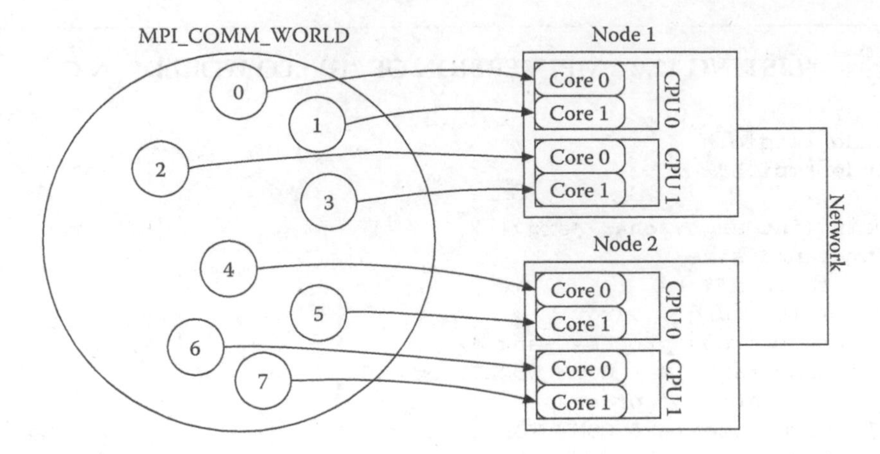

FIGURE 12.1
Relation between MPI ranks and cores.

or many cores of a central processing unit (CPU). Figure 12.1 describes an example of mapping from rank to cores.

At the end, MPI_Finalize is used to terminate the MPI execution environment.

```
int MPI_Finalize( void )
```

There are few exceptions, but you need to remember that nearly all MPI function calls need to be done after *MPI_Init* and before *MPI_Finalize.* Please note that it is common to see *MPI_Init* and *MPI_Finalize* in the main function, but they are not mandatory.

12.2.3 MPI_Comm_size and MPI_Comm_rank

In the previous section, we discussed the default communicator called *MPI_COMM_WORLD.* *MPI_Comm_size* and *MPI_Comm_rank* are two functions that are very common in MPI code, and these functions allow you to get more information regarding the communicator.

The C syntax of these two functions is as follows:

```
int MPI_Comm_size( MPI_Comm comm, int *size )
int MPI_Comm_rank( MPI_Comm comm, int *rank )
```

For these functions, the first parameter is a communicator and has the type defined by MPI for communicators, which is *MPI_Comm.* The second parameter is an output one. It returns the number of processes in the communicator for *MPI_Comm_size* and the rank (i.e., number) of the calling process in the communicator.

12.2.4 Compilation and Execution

Compilation and the way to execute the MPI code are slightly different compared to serial code. As shown in Figure 12.2, to compile a C code, you will usually need to use the

Language	Command
C	mpicc
C++	mpic++
Fortran	mpif90

FIGURE 12.2
Corresponding compilation commands for each language.

command mpicc. I say usually because the name of the command can differ depending of your MPI library. It is important to check the name of the different wrapper in the documentation of the MPI library that you are using. mpicc is a script and also a wrapper for the C compiler. This wrapper is always the same one whatever the compiler used (GNU, Intel, etc.). Why do we need that? To simplify the life of users. In fact, if I take the example of using the MVAPICH2 MPI implementation on the Stampede supercomputer at the Texas Advanced Computing Center, when I want to compile a C code:

```
$ mpicc test.c -o test
```

The wrapper is in reality doing:

```
icc -Wl,-rpath,/opt/apps/intel/15/composer_xe_2015.2.164/compiler/
lib/intel64 -Wl,-rpath,/opt/apps/intel/15/composer_xe_2015.2.164/
compiler/lib/intel64 -I/opt/apps/intel15/mvapich2/2.1/include
-L/opt/apps/intel15/mvapich2/2.1/lib -Wl,-rpath
-Wl,/opt/apps/intel15/mvapich2/2.1/lib-Wl,--enable-new-dtags -lmpi
test.c -o test
```

As you can see, using mpicc is easier than writing the complete command line. For other languages, the name of the wrapper is different. Figure 12.2 shows you the different name of the wrapper used for the different languages.

Once your code is compiled, you will need to execute it. To execute an MPI code, you will need to use a launcher. The launcher is taking care of the propagation of the code and the establishment of connections between nodes that are involved. Most of the time, each MPI library has its own launcher (mpirun_rsh for MVAPICH2, mpiexec.hydra for MPICH and Intel MPI, and orterun for Open MPI), but there is a common way to call them: by using mpirun. mpirun is a command commonly used by MPI libraries to call their own launcher. To launch four MPI tasks on two nodes (*node1* and *node2*), you will need to use the following command:

```
$ mpirun -np 4 -hostfile hosts ./my_code
```

The number after "-np" indicates the number of MPI tasks that you want. A file that contains the list of the hosts follows the command "-hostfile"; in our example, the file is called "hosts" and contains the following list:

```
node1
node1
node2
node2
```

This will inform the launcher to launch the two first MPI tasks on *node1* and the two last ones on *node2*. Be aware again here that the way to describe your list of nodes inside this file could be different. In fact, each MPI library has its own way to describe the number of tasks that it requires on the nodes. The version here is the generic one. If we take the initial hello_world example that we discussed earlier, the output of the file submitted through this command will be as follows:

```
Hello world from process 0 of 4
Hello world from process 1 of 4
Hello world from process 3 of 4
Hello world from process 2 of 4
```

You can notice two things:

- There are four processes, but the first one is number 0.
- The order is arbitrary.

These observations are correct. In fact, as mentioned earlier, the first MPI process in the communicator (provided by *MPI_Comm_rank*) has the number 0, and as there are four MPI tasks (provided by *MPI_Comm_size*), the last one has the number 3. Regarding the order of the output, this result is also expected. There is no automatic mechanism inside MPI to order your output. If you want to order it, you will have to create a code for that. To avoid this kind of issue, many programs only provide output from rank 0. We will see in Section 12.5 how to proceed to do that.

In this section, we saw some basic MPI functions and discovered how to compile and run an MPI program. However, inside MPI, there is message passing, that is, exchange of messages. In the next sections, we will see how to use the point-to-point messages and collective communication inside MPI.

12.3 Point-to-Point Communication

Point-to-point communication involves one sender and one receiver, that is, two MPI tasks. There are multiple modes of communications, but as this is mainly an introduction to MPI, we will only look at blocking and nonblocking point-to-point communication.

12.3.1 MPI_Send and MPI_Recv

MPI_Send is a blocking send. It means that the routine is blocked until it is safe to modify the application buffer for reuse. *MPI_Recv* is a blocking receive. It returns only after the data arrived and are ready to be used by the program.

The parameters of *MPI_Send* are the following:

```
MPI_Send(const void *buf, int count, MPI_Datatype datatype, int
dest, int tag, MPI_Comm comm)
```

And the parameters of *MPI_Recv* are the as follows:

```
MPI_Recv(void *buf, int count, MPI_Datatype datatype, int source,
int tag, MPI_Comm comm, MPI_Status *status)
```

There are few parameters, which are common or specific to a function. Let's look at them:

- **buf* is the initial address of the send or receive buffer.
- *count* contains the number of items to send or receive.
- *datatype* is the MPI data type of item that you want to receive or send.
- *tag* is like the message ID. You can only receive a message that matches your tag.
- *comm* is the communicator, and of course, both sender and receiver need to be in the same communicator.
- *status* returns information on the message received. It indicates the source and the tag of the message.
- *dest* is the rank of the destination.
- *source* is the rank of the source.

These parameters are perhaps may not be very clear, but we will see how they work exactly in an example. First though, I would like to raise a point. As you can see, there is a parameter for the MPI datatype. A classical question is: Why do I need a specific datatype when there is one that already exists in C or Fortran?

The answer is simple: MPI is designed to be able to run on heterogeneous systems, and sometimes, you can have 64-bit nodes exchanging messages with 32-bit nodes. The fact that MPI has its own datatype allows the MPI library to speak with all these different systems in a common language. The MPI library takes care of the translation to convert the value in a different system.

Table 12.1 provides an example of corresponding datatypes between MPI and language C. This point is important, and it illustrates that MPI aims to be portable.

Now, let's look at an example using *MPI_Send* and *MPI_Recv* in Listing 12.3.

LISTING 12.3 EXAMPLE USING MPI_Send AND MPI_Recv

```
#include <stdio.h>
#include "mpi.h"

int main (int argc, char *argv[]){
    int rank, number;
    number =0;
    /* starts MPI */
    MPI_Init (&argc, &argv);
    /* get current process rank */
    MPI_Comm_rank (MPI_COMM_WORLD, &rank);
    if (rank == 0) {
        number = 123;
        MPI_Send(&number, 1, MPI_INT, 1, 0, MPI_COMM_WORLD);
    } else if (rank == 1) {
        printf("Number = %d\n", number);
        MPI_Recv(&number, 1, MPI_INT, 0, 0, MPI_COMM_WORLD,
MPI_STATUS_IGNORE);
        printf("Process 1 received data from process 0, number =  %d
\n", number);
    }
    MPI_Finalize(); /* ends MPI */
    return 0;
}
```

This example is simple and will help you understand how to use *MPI_Send* and *MPI_Recv.* The following shows how the code proceeds. You have to remember that the same code will be executed by all MPI tasks:

- At the beginning, we initialize the MPI code using *MPI_Init.*
- Then, we ask for the rank of the MPI process by using *MPI_Comm_rank.*

TABLE 12.1

Corresponding Datatypes between MPI and Language C

MPI Datatype	MPI Datatype C Equivalent
MPI_SHORT	short int
MPI_INT	int
MPI_LONG	long int
MPI_LONG_LONG	long long int
MPI_UNSIGNED_CHAR	unsigned char
MPI_UNSIGNED_SHORT	unsigned short int
MPI_UNSIGNED	unsigned int
MPI_UNSIGNED_LONG	unsigned long int
MPI_UNSIGNED_LONG_LONG	unsigned long long int
MPI_FLOAT	float
MPI_DOUBLE	double
MPI_LONG_DOUBLE	long double
MPI_BYTE	unsigned char

- If the process is rank 0, the value of the number is changed to 123, then the process 0 sends this number (only one element), which is an integer, to process 1 with the tag 0 using the communicator *MPI_COMM_WORLD*.
- Process 1 prints out the value of the number then overwrites this value with the integer received by the message from rank 0 with the tag 0 using *MPI_COMM_WORLD*. Notice that *MPI_Recv* also used the macro *MPI_STATUS_IGNORE* to mention that it was ignoring the status, then rank 1 prints out the new value.

Here is the output of this code:

```
Number = 0 Process 1 received data from process 0, number = 123
```

As you can see, rank 0 exchanges a message with rank 1, which changes the value of number. In your opinion, what will be the result with 5 or 128 MPI tasks? It will be exactly the same, because we are using a condition branch that involves only rank 0 and 1.

12.3.2 MPI_Isend and MPI_Irecv

Another way to send messages is by using nonblocking communications (*MPI_Isend* and *MPI_Irecv*); here, the routines will return almost immediately. The big advantage here is that you can do computation while the MPI library is trying to send a message. It is called overlapping.

Nonblocking send is defined as below:

```
MPI_Isend(void *buf, int count, MPI_Datatype dType, int dest, int
tag,  MPI_Comm comm, MPI_Request *request)
```

And the parameters of MPI_Irecv are the following:

```
MPI_Irecv(void *buf, int count, MPI_Datatype dType, int source, int
tag, MPI_Comm comm, MPI_Request *request)
```

You must be careful because it is unsafe to modify the application buffer until you know for a fact that the requested nonblocking operation was performed. How can you know this? By using functions like *MPI_Test* and *MPI_Wait*. Before providing more details, let's look at the syntax of these functions:

```
MPI_Test(MPI_Request *request, int *flag, MPI_Status *status)
MPI_Wait(MPI_Request *request, MPI_Status *status)
```

MPI_Request objects are used by nonblocking send and receive calls to know the current state of the communication (in progress or completed). *MPI_Wait* is a blocking routine, and it blocks until a specified nonblocking send or receive operation has been completed. On the other side, MPI_Test tests for the completion of a request and returns with flag marked true if the operation is completed.

Understanding this can be a bit complicated without an example. Listing 12.4 does the same thing that the previous example did, but instead of using a blocking operation, it will use nonblocking ones.

LISTING 12.4 EXAMPLE USING MPI_Isend AND MPI_Irecv

```
#include <stdio.h>
#include "mpi.h"

int main (int argc, char *argv[]){
    int rank, source, destination, number;
    number=0;
    source=0;
    destination=1;
    MPI_Status status;
    MPI_Request request;

    MPI_Init (&argc, &argv); /* starts MPI */
    MPI_Comm_rank (MPI_COMM_WORLD, &rank); /* get current process
rank */
    request=MPI_REQUEST_NULL;
    if (rank == 0) {
        number = 123;
       MPI_Isend(&number,1,MPI_INT,destination,0,MPI_COMM_WORLD,
&request);
    } else if (rank == 1) {
        MPI_Irecv(&number, 1, MPI_INT, source, 0, MPI_COMM_WORLD,
&request);
    }
    MPI_Wait(&request,&status);
    if(rank == source){
      printf("processor %d  sent %d\n", rank, number);
    }
    if(rank == destination){
      printf("processor %d  got %d\n", rank, number);
    }

    MPI_Finalize(); /* ends MPI */
    return 0;
}
```

Here, rank 0 will post the nonblocking send, rank 1 will post the nonblocking receive, then the code will reach *MPI_Wait*, which will block the execution until the end of the exchange between the two MPI processes. The final output will be exactly the same as in the version with blocking send and receive.

12.3.3 Potential Pitfalls

There are few points where you need to be careful to avoid issues, such as hangs or deadlocks with point-to-point operations. First, be sure that the value of tag is the same on the receiving and sending sides. A solution to avoid this problem is to use the macro *MPI_ANY_TAG* as tag. It will allow the receiver to accept a message with any tag value. But be sure that your MPI tag does not receive multiple messages with different values before or it could lead to wrong results. Second, a common issue is the deadlock.

LISTING 12.5 EXAMPLE USING MPI_Isend AND MPI_Irecv

```
  if (rank == 0) {
  dest = 1;
  source = 1;
  MPI_Recv(&inmsg, 1, MPI_CHAR, source, tag, MPI_COMM_WORLD, &Stat);
  MPI_Send(&outmsg, 1, MPI_CHAR, dest, tag, MPI_COMM_WORLD);
}else if (rank == 1) {
  dest = 0;
  source = 0;
  MPI_Recv(&inmsg, 1, MPI_CHAR, source, tag, MPI_COMM_WORLD, &Stat);
  MPI_Send(&outmsg, 1, MPI_CHAR, dest, tag, MPI_COMM_WORLD);
}
```

If you look at Listing 12.5, it will deadlock. Why? Because if you look carefully, you will see that both MPI tasks are waiting for a message, and as it is a blocking receive, the program will be blocked at this state until the reception of the message. Unfortunately, the send is done after the reception. So, your code will hang until you stop it. The order of your message is very important, and you have to be careful with it. How do you fix it? You just have to change the order on one side, as described in Listing 12.6.

LISTING 12.6 EXAMPLE USING MPI_Isend AND MPI_Irecv

```
  if (rank == 0) {
  dest = 1;
  source = 1;
  MPI_Recv(&inmsg, 1, MPI_CHAR, source, tag, MPI_COMM_WORLD, &Stat);
  MPI_Send(&outmsg, 1, MPI_CHAR, dest, tag, MPI_COMM_WORLD);
}else if (rank == 1) {
  dest = 0;
  source = 0;
  MPI_Send(&outmsg, 1, MPI_CHAR, dest, tag, MPI_COMM_WORLD);
  MPI_Recv(&inmsg, 1, MPI_CHAR, source, tag, MPI_COMM_WORLD, &Stat);
}
```

Now, everything will be fine.

12.4 Collective Communication

Collective communications are used when you need to communicate with more than two MPI processes. There are different kinds of schemes for collectives: one-to-many, many-to-one, and many-to-many.

Since this is just an introduction to MPI, we will only describe four MPI collectives that are generally used inside MPI code: *MPI_Bcast, MPI_Reduce, MPI_Scatter,* and *MPI_Gather.*

It is important to note that all processes within the communicator group call the same collective communication function with matching arguments. And just like point-to-point, the size of the data sent must exactly match the size of the data received. The aim of this section is to explain the aim of collectives by describing a few of them. The next section shows an example using collectives, but if you are really interested in learning more about collectives, I invite you to look at the recommended books mentioned in the conclusion.

12.4.1 MPI_Bcast

```
MPI_Bcast( void *buffer, int count, MPI_Datatype datatype, int root,
MPI_Comm comm )
```

MPI_Bcast broadcasts (sends) the content of the buffer (*buffer*) from the process designated as *"root"* to all other processes in the group. The message is identical for all destinations and it can contain one or more elements. (It will depend on the value of count.)

12.4.2 MPI_Reduce

```
MPI_Reduce(const void *sendbuf, void *recvbuf, int count, MPI_
Datatype datatype, MPI_Op op, int root, MPI_Comm comm)
```

MPI_Reduce applies a reduction operation (*op*) on all tasks in the communicator (*comm*) and places the result in the buffer (*recvbuf*) of the root task.

MPI has predefined multiple operations for *MPI_Reduce*. Figure 12.3 provides a list of some of the operations.

MPI Reduction Operation	Description
MPI_MAX	Maximum
MPI_MIN	Minimum
MPI_SUM	Sum
MPI_PROD	Product

FIGURE 12.3
A list of some of the operations available with MPI_Reduce.

12.4.3 MPI_Scatter

```
MPI_Scatter(const void *sendbuf, int sendcount, MPI_Datatype
sendtype, void *recvbuf, int recvcount, MPI_Datatype recvtype, int
root, MPI_Comm comm)
```

Contrary to *MPI_Bcast, MPI_Scatter* distributes distinct messages from a single source task (*sendbuf*) to each task (*recvbuf*) in the communicator. If you want, for example, to distribute the different components of a vector to different MPI tasks (like for Matrix multiplication), *MPI_Scatter* is the function to use.

12.4.4 MPI_Gather

```
MPI_Gather(const void *sendbuf, int sendcount, MPI_Datatype
sendtype, void *recvbuf, int recvcount, MPI_Datatype recvtype, int
root, MPI_Comm comm)
```

MPI_Gather is the reverse of *MPI_Scatter* and gathers distinct messages from each task (*sendbuf*) in the communicator (*comm*) to a single destination task (*recvbuf*). Be aware that the *recvcount* parameter is the count of elements received per process, not the total summation of counts from all the processes.

12.5 Example: Compute π

In the previous section, collective communications were described, but no real example was provided to show you how to use them. Listing 12.7 is a complete MPI code which aims to compute the value of π before, to compare the result with the reference value of π.

LISTING 12.7 π COMPUTATION WITH MPI

```
#include <stdio.h>
#include <stdlib.h>
#include <math.h>
#include "mpi.h"
```

```
int main(int argc, char *argv[])
{
    int true=1; /* to force the loop */
    int  n, /* number of intervals */
         rank, /* rank of the MPI process */
         size, /* number of processes */
         i, /* variable for the loop */
         len;  /* name of the process */
    double PI_VALUE = 3.14159265358979323846264 3;
    /* reference value of pi */
    double mypi, /* value for each process */
           pi, /* value of pi calculated */
           h,
           sum, /*sum of the area */
           x;

    double start_time, /* starting time */
           end_time, /* ending time */
           /* time for computing value of pi */
           computation_time;

    char name[80]; /* char array for storing the name of
                      the node where is located the
                      process */

    /*Initialization */
    MPI_Init(&argc,&argv);
    MPI_Comm_size(MPI_COMM_WORLD,&size);
    MPI_Comm_rank(MPI_COMM_WORLD,&rank);

    /* We ask for the name of the node */
    MPI_Get_processor_name(name, &len);

    while (true)
    {
    if (rank == 0) {
        printf("Enter the number of intervals: (0 quits)");
        scanf("%d",&n);
        start_time = MPI_Wtime();
    }
    /* We are broadcasting to everybody the number of interval */
    MPI_Bcast(&n, 1, MPI_INT, 0, MPI_COMM_WORLD);

    if (n == 0) break;

    h   = 1.0 / (double) n;
    sum = 0.0;
    for (i = rank + 1; i <= n; i += size) {
        x = h * ((double)i - 0.5);
        sum += 4.0 / (1.0 + x*x);
    }
    mypi = h * sum;
```

```
    printf("This is my sum: %.16f from rank: %d in: %s\n", mypi,
           rank, name);

    MPI_Reduce(&mypi, &pi, 1, MPI_DOUBLE, MPI_SUM, 0, MPI_COMM_WORLD);

    if (rank == 0){
        printf("pi is approximately %.16f, Error is %.16f\n", pi,
               fabs(pi - PI_VALUE));
        end_time = MPI_Wtime();
        computation_time = end_time - start_time;
        printf("Time of calculating pi is: %f\n", computation_time);
    }
    }
    MPI_Finalize();
    return 0;
}
```

The code can appear complex, but you will see that it is very simple. At the beginning, a bunch of variables are defined for the different functions that we will use. Then, we have the classical *MPI_Init, MPI_Comm_size,* and *MPI_Comm_rank* to get all initial information. After, we call an MPI function that we did not see earlier: *MPI_Get_proccessor_name.* This function is used mainly to know the name of the node where the MPI rank is. Please remember that each MPI rank has its own memory space and is executing the same code in its own space.

Then, we enter into a loop, and rank 0 requests the number of intervals that you want to use for the computation. Once you enter the number, we start to count the time by using MPI_Wtime. This function returns the time in seconds.

After, rank 0 broadcasts to all MPI processes the number of intervals (value of n). Then, each MPI process computes its own part before the call to *MPI_Reduce,* which will sum all the different parts before writing the result in the buffer, called π of rank 0. Once this is done, rank 0 prints out the value computed, shows the error compared to the real value, and provides the time that it took to do the whole operation. In this code, the root process for the collectives was rank 0. This case is very common in MPI code. Another point to raise is that all MPI tasks were used to do the computation. You can try to increase the number of intervals and the number of MPI tasks and you will see that the time will change based on these two elements. The figure below is an example of output with 16 MPI tasks using 8 MPI tasks per node.

```
    $ mpirun_rsh -np 16 -hostfile hosts ./a.out
Enter the number of intervals: (0 quits)1000
This is my sum: 0.1982883627376218 from rank: 0 in: node1
This is my sum: 0.1981632434099282 from rank: 1 in: node1
This is my sum: 0.1980379960196081 from rank: 2 in: node1
This is my sum: 0.1979126211916232 from rank: 3 in: node1
This is my sum: 0.1977871195509662 from rank: 4 in: node1
This is my sum: 0.1976614917226540 from rank: 5 in: node1
This is my sum: 0.1975357383317214 from rank: 6 in: node1
This is my sum: 0.1974098600032138 from rank: 7 in: node1
```

```
This is my sum: 0.1952848571121811 from rank: 8 in: node2
This is my sum: 0.1951607287836729 from rank: 9 in: node2
This is my sum: 0.1950364753927391 from rank: 10 in: node2
This is my sum: 0.1949120975644254 from rank: 11 in: node2
This is my sum: 0.1947875959237667 from rank: 12 in: node2
This is my sum: 0.1946629710957804 from rank: 13 in: node2
This is my sum: 0.1945382237054593 from rank: 14 in: node2
This is my sum: 0.1944133543777651 from rank: 15 in: node2
pi is approximately 3.1415927369231262, Error is 0.000000083
3333331
Time of calculating pi is: 0.005457
```

If you keep the same interval and increase the number of MPI tasks, you may be surprised, but I will not explain why. I will let you try to understand and think about what's going on.

12.6 Conclusion

Providing an introduction to MPI in a chapter is a complicated task. To learn MPI, you need a complete book, not a chapter. Throughout this chapter, I have tried to provide you with a global overview of MPI, which should be some basic knowledge. Some basic MPI functions were given before learning how to compile and execute an MPI. The two last points are very important, and it is important to review the documentation of the MPI library that you are using to ensure that you are doing everything correctly. Using the launcher with a wrong command or wrong description of the hostfile can really impact the performance of your MPI code. We then discussed the different ways to exchange messages inside MPI by using point-to-point and collective communications. These sections were short introductions, and to learn more about MPI, I recommend the following books:

- *Using Advanced MPI: Modern Features of the Message Passing Interface* by William Gropp, Torsten Hoefler, Rajeev Thakur, and Ewing Lusk
- *Using MPI: Portable Parallel Programming with the Message Passing Interface,* 2nd Edition, by William Gropp, Ewing Lusk, and Anthony Skjellum

Both are very good and explain MPI and all the new features introduced with the last version of the standard in detail. Some final advice regarding MPI: please consult the user guide if you are running MPI codes on a large installation, as such systems may have their own caveats with regard to how codes are run.

13

Introduction to OpenMP

Yaakoub El Khamra

CONTENTS

13.1 What Is OpenMP?

Open Multi-Processing (OpenMP) [1] is a set of compiler directives, environment variables, and a runtime library that extend C, C++, and Fortran languages to express shared memory parallelism or multithreading. OpenMP allows users to write code that "forks" a team of threads and distributes the workload among them, thereby reducing the overall time to completion. The number of threads is usually equal to the number of cores available on the system such that each thread runs on its own core and all cores utilized. Server and workstation processor design trends in the past 15 years have been focused on adding more cores per chip. Multithreading using OpenMP offers a convenient, scalable means of improving application performance especially with the higher core count on a single machine.

In this chapter, we will introduce OpenMP and highlight its strengths and weaknesses. The OpenMP statements commonly used to add parallelism to typical scientific applications will also be introduced. By the end of this chapter, the reader is expected to be able to parallelize C/C++ or Fortran code and solve typical parallelization errors such as race conditions.

13.2 OpenMP and Multithreading

Other programming platforms and libraries that use multithreading include pthreads (POSIX threads), CoArray Fortran, TBB (Thread Building Blocks [2]), and Cilk [3]. Pthreads relies on manual synchronization objects (mutex and semaphore objects) in C/C++, making them difficult to use in Fortran. Furthermore, programming at this level of fine grain thread control makes pthreads applications difficult to implement and debug. CoArray Fortran is, as the name suggests, limited to certain Fortran compilers but offers high-level controls over parallelism. TBB, the Thread Building Blocks library, offers the same high-level control over threads but is C++ only. Cilk, on the other hand, is easy to learn and offers high-level controls but is its own C-like programming language. In summary, it is fair to say that OpenMP is not unique in its approach to parallelism (i.e., multithreading); it is, however, supported on most architectures, by most compilers, and is more widely used in scientific applications.

The advantages of using OpenMP over other parallelism methodologies include the following:

- *Portability*: OpenMP is supported by open-source and commercial compilers on Unix-like environments and Windows.
- *Efficiency*: OpenMP threads are lightweight and implementations have low overhead.
- *Ease of use*: Incrementally adding OpenMP parallelism to existing codes, especially loop intensive scientific codes, is straightforward and unobtrusive.
- *Control over the level of parallelism*: Coarsely, threads are created at application start and exist throughout. Finely, threads are created and destroyed in the intensive (aka "hot") loops.
- *Prevalence of multicore hardware*: Cell phones, entry-level laptops, accelerators (graphics processing units, Many Integrated Core [MIC] cards), and of course, supercomputers all have multiple processing cores in the same processor, often multiple processors.

There are, however, some limitations:

- OpenMP will scale but is limited in scalability by the memory architecture.
- Overhaul of a large and complex serial code to OpenMP can be challenging.

From a systems point of view, the application is one process with many threads. Each thread has access to the entire memory allocated to the process and is given some private memory that is available only to that thread. The concurrent execution of computational work (or tasks) using the shared and private memory is the definition of parallel computing. Tasks execute independently and synchronize through barriers. Task updates to variables in shared memory should be mutually exclusive; otherwise, subtle errors known as race conditions can arise.

13.3 Compiling Applications with OpenMP (GCC, Intel, PGI), Fortran, and C

Compiling applications with OpenMP support is straightforward from the command line on Unix/Linux and BSD systems. No special hardware is required; however, for the highest performance, a high-core-count (and high-bandwidth) machine would be recommended. Most server/workstation grade systems have more than eight cores today and most laptops have four or better. Given a compiler, the suitable compiler flag can be found in the following:

GNU [4] (gcc,g++,gfortran)	-fopenmp
Intel (icc,icpc,ifort)	-openmp
Portland Group (pgcc,pgCC,pgf77,pgf90)	-mp

On Windows, the flags are slightly different:

GNU (gcc,g++,gfortran)	-fopenmp
Intel (icc,icpc,ifort)	/Qopenmp
MS Visual Studio	/openmp

The following examples in C and Fortran 90 are complete OpenMP applications that print the message "Hello World" from different threads and provide the thread count.

HelloWorld.c C example:

```
#include<stdio.h>
#include<omp.h>

int main(int argc, char ** argv)
{
  #pragma omp parallel
  {
    int thread_id = omp_get_thread_num();
    int thread_count = omp_get_num_threads();
    printf("Hello from thread number: %d out of: %d\n",
     thread_id, thread_count);
  }
  return 0;
}
```

HelloWorld.f90 Fortran 90 example:

```
program main

use omp_lib
implicit none

integer thread_id, thread_count

!$omp parallel private(thread_id,thread_count)
thread_id = omp_get_thread_num()
thread_count = omp_get_num_threads()
write (*,*) "Hello from thread number:",thread_id, "out of:
",thread_count
!$omp end parallel
end program main
```

To compile these simple "Hello World" type applications on Linux,

```
$ gcc -fopenmp HelloWorld.c -o HelloWorld_c
```

And to run it:

```
$ ./HelloWorld_c
```

The output from this application on a four core machine would be as follows:

```
Hello from thread number: 0 out of: 4
Hello from thread number: 1 out of: 4
Hello from thread number: 3 out of: 4
Hello from thread number: 2 out of: 4
```

Note that thread execution is generally out of order; hence, the output to the screen is not ordered by the thread id. For the Fortran 90 version:

```
$ gfortran -fopenmp HelloWorld.f90 -o HelloWorld_f90
$ ./HelloWorld_f90
```

13.4 OpenMP Basic Environment Variables

Environment variables provide easy, runtime controls to the behavior of OpenMP code. Perhaps, the most important environment variable is OMP_NUM_THREADS, which sets the number of threads to use during execution. Setting this environment variable on the command line can be done as follows:

Bash:

```
$ export OMP_NUM_THREADS=4
```

And csh:

```
$ setenv OMP_NUM_THREADS=4
```

Setting OMP _ NUM _ THREADS to 4 will cause all applications running in the same command line to execute with four threads. The number of threads can also be modified in code through OpenMP directives or even OpenMP functions (more on this later). It is worthwhile to experiment with the OMP _ NUM _ THREADS environment variable and the simple "HelloWorld" examples. Keep in mind that the number of threads used to run the application is not necessarily limited by the number of cores. Oversubscribing the cores (more than one thread per core) is entirely possible and the application would run correctly, only slower. We will discuss other environment variables later on in this chapter.

13.5 OpenMP Basics

13.5.1 Syntax, Directives, and Clauses (Examples in C and Fortran)

The range of OpenMP constructs include compiler directives and the runtime environment (function calls, environment variables). Figure 13.1 illustrates the range of possible constructs.

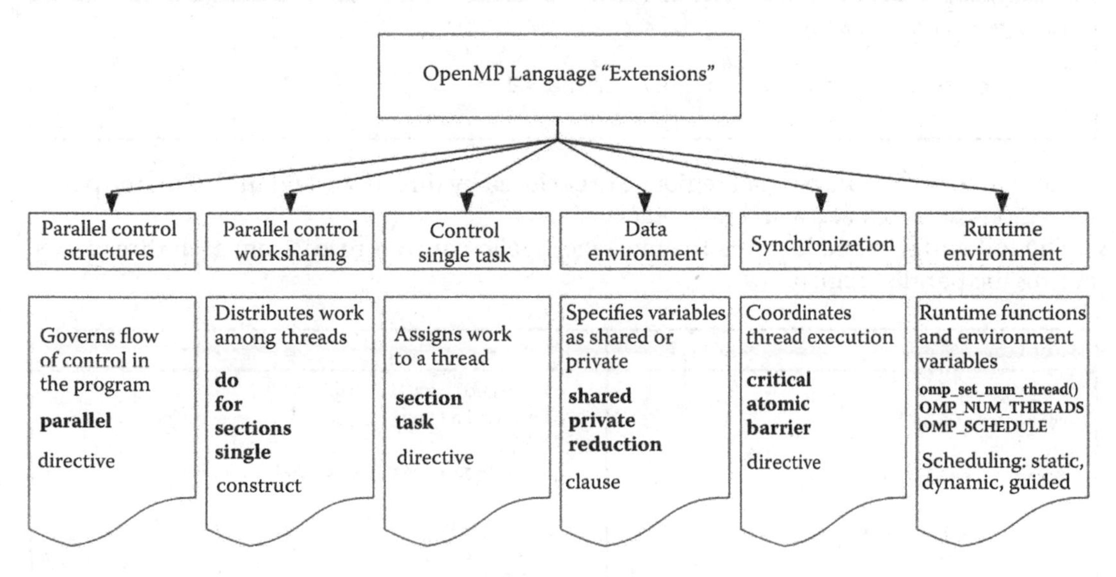

FIGURE 13.1
OpenMP constructs that are currently available in all modern OpenMP implementations.

The "language" uses compiler directives (not to be confused with compiler flags), which programmers have to write in the source code. These directives are treated as comments if the compiler is not given the necessary OpenMP flag. OpenMP compiler directives are classified into three main categories: the sentinel, construct, and clause(s).

```
OpenMP Directives:  Sentinel, construct and clauses
        #pragma omp  construct [clause [[,]clause]…]       C/C++
        !$omp        construct [clause [[,]clause]…]       F90
```

For example:

```
        #pragma omp        parallel num_threads(4)          C/C++
        !$omp      parallel num_threads(4)                  F90
```

Function prototypes and types are in the file:

```
        #include <omp.h>                                    C/C++
        use omp_lib                                         F90
```

Most OpenMP constructs apply to a "structured block," that is, a block of one or more statements with one point of entry at the top and one point of exit at the bottom.

A good example of an OpenMP directive is the directive to create a team of threads to execute sections of code in parallel. In Fortran:

```
!$OMP parallel
  ...   < perform statements here in parallel >
!$OMP end parallel
```

and in C/C++:

```
#pragma omp parallel
{
...   < perform statements here in parallel >
}
```

Note that in Fortran, parallel regions are enclosed by directives, and in C/C++, the parallel regions are enclosed with curly brackets.

The following example prints the message "Hello I am a thread" from each thread executing the parallel region.

Fortran

```
program main
implicit none

!$omp parallel
write (*,*) "Hello I am a thread"
!$omp end parallel

end program main
```

C/C++

```
#include<stdio.h>
int main(int argc, char ** argv)
{
  #pragma omp parallel
  {
    printf("Hello I am a thread\n");
  }
  return 0;
}
```

The output printed to the screen would be:

```
$ export OMP_NUM_THREADS=5
$ ./Hello_I_am_a_Thread
Hello I am a thread
Hello I am a thread
Hello I am a thread
Hello I am a thread
Hello I am a thread
```

Note that in this case, it is repeated five times because `OMP _ NUM _ THREADS` was set to 5 before the application (`Hello _ I _ am _ a _ Thread`) was run.

13.6 Parallel Regions and Worksharing

13.6.1 OpenMP Do/For

All programs begin as a single process, a master thread that executes in serial mode until a parallel region construct is encountered. The master thread then creates (forks) a team of parallel threads that simultaneously execute tasks in a parallel region. After executing the statements in the parallel region, the team threads synchronize and terminate (join) and the master thread continues executing the rest of the application. Figure 13.2 shows this behavior for four threads.

It is important to note the implied barrier at the synchronization before threads merge.

To distribute the workload across the threads created in the parallel region, one would need to use worksharing constructs. These include constructs to distribute iterations of a for/do loop:

Fortran	C/C++

`!$omp parallel`
`!$omp do`
`do i=1,n`
`   call work(i)`
`end do`
`!$omp end do`
`!$omp end parallel`
	`#pragma omp parallel`
	`{`
	`#pragma omp for`
	`  for(i=0;i<n;i++)`
	`  {`
	`    work(i)`
	`  }`
	`}`

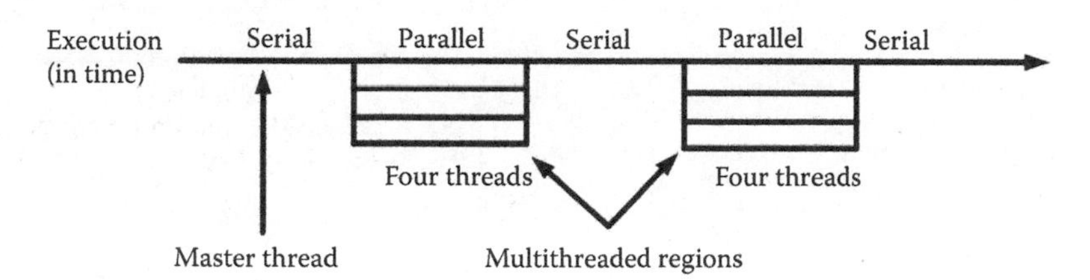

FIGURE 13.2
Multithreaded regions separated by serial execution. Teams of threads (four threads each in this instance) are forked and joined at least twice in this application.

In the above example, the do/for loop iterations are split among the threads via the do/for worksharing construct. These two directives (one for parallel region, the other for iteration) can be combined into a single construct for brevity as follows:

Fortran	C/C++

!$omp parallel do
do i=1,n
   call work(i)
end do
!$omp end parallel do
	#pragma omp parallel for
	for(i=0;i<n;i++)
	  {
	    work(i)
	  }

Note the absence of curly brackets { } in the C/C++ version of the worksharing construct when we combine the `parallel` and `for` constructs. Furthermore, having an iteration counter (enumerator) is necessary for parallelization. Skipping iterations with `cycle/continue` statements is handled properly by OpenMP. Early loop exit with a `break` or `exit` statement is not supported. An example illustrating work distribution, and in a couple of cases replication, is shown in Figure 13.3.

In the leftmost case, assuming the application is run with four threads, the same exact "`code`" is executed on all four threads. In the middle case, the iterations 1 to N*4 are distributed. Everything being equal, each of the four threads will execute exactly N iterations. In the rightmost case, a mixture of replication and worksharing occurs: "`code1`" is replicated, "`code2`" is shared, and "`code3`" is replicated.

It is important to note that parallel regions containing multiple workshares can be merged. Merging worksharing constructs into a single parallel region eliminates the

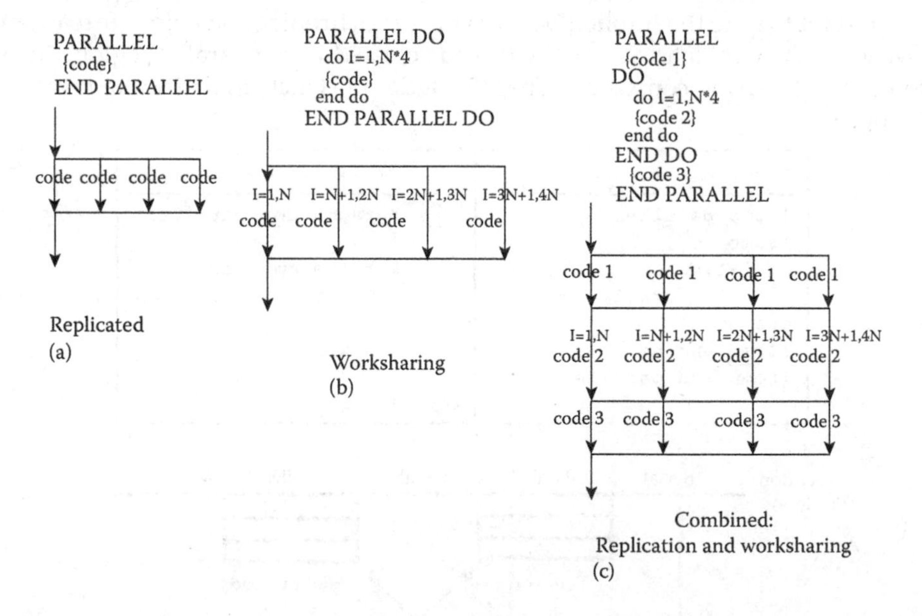

FIGURE 13.3
(a) A parallel section without any work distribution, that is, work replication. (b) A parallel section with work distribution. (c) A combination of work replication, distribution, and replication.

overhead of separate team formations. This overhead is usually small, but the accumulated cost of thread creation and deletion can get high if it is performed repeatedly (inside a bigger loop for example). The following example illustrates this point.

Separate parallel regions	Merged parallel regions

`!$omp parallel do`
`      do i=1,n`
`         a(i)=b(i)+c(i)`
`      enddo`
`!$omp end parallel do`
 	
`!$omp parallel do`
`      do i=1,n`
`         x(i)=y(i)+z(i)`
`      enddo`
`!$omp end parallel do`
	`!$omp parallel`
	`  !$omp do`
	`      do i=1,n`
	`         a(i)=b(i)+c(i)`
	`      enddo`
	`  !$omp end do`
	
	`  !$omp do`
	`      do i=1,n`
	`         x(i)=y(i)+z(i)`
	`      enddo`
	`  !$omp end do`
	`!$omp end parallel`

A quick distinction needs to be made between lexical and dynamic scoping for parallel regions.

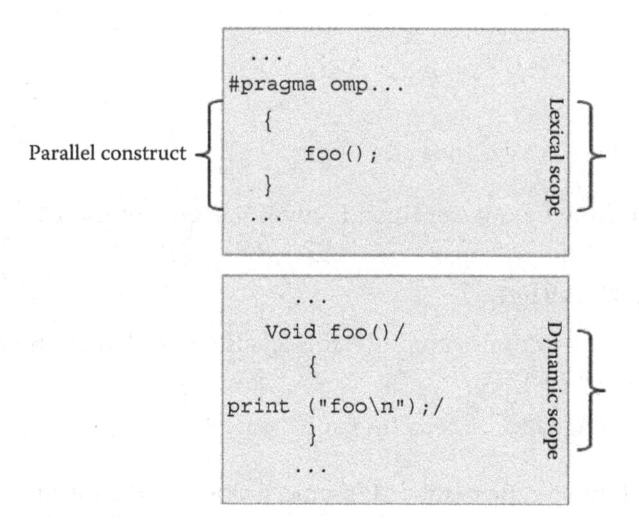

Worksharing directives that appear outside the lexical extent of a parallel directive are called "orphaned worksharing constructs." The lexical extent is also called the "parallel construct." When in the dynamic extent (i.e., within a called function or subroutine in a parallel region), the worksharing construct behavior is identical (almost) to a worksharing construct within the parallel region: variables take on the share/private attributes of the callee.

Fortran	C/C++

`!$omp parallel`
`   call work(n,a,b,c)`
`!$omp end parallel`
 	
`subroutine work(n,a,b,c)`
`use omp_lib`
`integer n, id, i`
`real*8, dimension(n):: a,b,c`
`print*,omp_get_num_threads();`
`!$omp do`
`   do i = 1,n`
`      a(i) = b(i) + c(i)`
`   end do`
`!$omp end do`
`end subroutine`
	`#pragma omp parallel`
	`{`
	`    work(n,a,b,c);`
	`}`
	`int work(int n, double *a,…)`
	`{`
	`int id, i;`
	`printf("%d",omp_get_num_threads());`
	`#pragma omp for`
	`   for(i=0; i<n; i++){`
	`      a[i] = b[i] + c[i];`
	`   }`
	`}`

When encountered from outside a parallel region (i.e., called from a serial portion of code), the master thread is the "team of threads." It is safely invoked as serial code.

OpenMP is efficient and forking threads have low overhead, as discussed earlier. There are, however, instances where the loops are so small that performing them sequentially can be computationally cheaper than creating threads, scheduling the iterations then deleting the threads. Luckily, the parallel construct supports an if clause that accepts a logical expression. For example:

```
!$omp parallel if (N>1000)
```

or

```
#pragma omp parallel if(N>1000)
```

Threads will fork from the master thread only if `N` is greater than `1000`.

13.6.2 Scheduling Constructs

The distribution of the iterations across the threads is controlled by the scheduling clause. The syntax would be as follows:

```
parallel do/for schedule(schedule-type[,chunk size])
```

There are four schedule types: static, dynamic, guided, and runtime.

- schedule (**static**, chunk): Threads receive chunks of iterations in thread order, round-robin, and the number of iterations is divided as equally as possible if no chunk size is specified. Static scheduling is good if every iteration contains the same amount of work and it helps keep parts of an array in the processor cache, making it very desirable for blocked array and stencil operations.
- schedule (**dynamic**, chunk): Threads receive iterations in chunks as the thread becomes available for more work. The workload is balanced across the threads, and this scheduling is best used with workloads that vary per iteration. There is, however, some overhead incurred.

- schedule (**guided**, chunk): Threads receive iterations in chunks as the thread becomes available for work, but the chunk size decreases exponentially until it reaches the chunk size specified. This scheduling is not only load balanced but also reduces the number of requests for more work.
- schedule (**runtime**): The scheduling in this case is determined at runtime by the environment variable OMP_SCHEDULE and is mostly used when experimenting with various scheduling clauses.

13.6.3 Sections and Tasks

Another type of worksharing construct, designed for use with noniterative algorithms, is sections. Sections are used to divide code into blocks that can be executed in parallel. Threads can execute more than one block or no blocks at all. Section constructs need to be enclosed in a parallel region to execute on different threads.

Fortran	C/C++
!$OMP PARALLEL SECTIONS	**#pragma omp parallel sections**
!$OMP SECTION	{
CALL sub1()	**#pragma omp section**
!$OMP SECTION	{ func1( ); }
CALL sub2()	**#pragma omp section**
!$OMP SECTION	{ func2 ( ); }
CALL sub3()	**#pragma omp section**
!$OMP END SECTIONS	{ func3 ( ); }
	}

In the above example, subroutines/functions 1 to 3 are executed on different threads but not necessarily in the order in which they appear in the code.

A more advanced worksharing construct with a similar purpose to sections is available in OpenMP: `task`. Tasks must be declared in a single construct so that only one thread creates the tasks, and tasks are only guaranteed to be evaluated at scheduling points such as `taskwait` or as barriers (whether implicit or explicit). As soon as one thread creates a task, it is sent to the task queue, where it can be evaluated by any idle thread within the parallel region. Tasks can be used to workshare recursive workloads such as searching through a tree or a linked list, but sections cannot.

#pragma omp single	only one thread should create tasks and add them to the task queue, implies a barrier at end
{	
#pragma omp task	a call to func1 is now added to the task queue
func1();	
#pragma omp task	a call to func2 is now added to the task queue
func2();	
}	implicit barrier at which the task queue should be empty and all tasks completed

13.6.4 Single and Master

The last worksharing constructs are "single" and "master." A block of code marked with a "single" worksharing construct will be executed by a single thread, and a block of code marked with "master" will be executed on the master thread. Only "single" has an implied barrier at the end. An example of its use is shown below.

Fortran	C/C++

`!$OMP single`
`glob_count = glob_count + 1`
`print *, glob_count`
`!$OMP end single`
	`#pragma single`
	`{`
	`glob_count++;`
	`printf("%d\n", glob_count);`
	`}`

13.7 Data Scoping

Data scoping clauses are used to specify how OpenMP handles variables in parallel regions. They follow directives, take a list of variables as arguments, and take the following form:

```
#pragma omp directive-name [clause(variable list)...]
!$omp directive-name [clause(variable list)...]
```

Commonly used data scoping clauses include the following:

- `private(variable list)`: Each thread has its own copy of the specified variable. Variables are undefined after worksharing region.
- `shared(variable list)`: Threads share a single copy of the specified variable.
- `default(type)`: A default of private, shared, or none can be specified. Note that loop counter(s) of worksharing constructs are always private by default; everything else is shared by default
- `firstprivate(variable list)`: Like private, but copies are initialized using value from the master thread's copy.
- `lastprivate(variable list)`: Like private, but the final value is copied out to the master thread's copy. With for/do workshares, the last iteration is copied, and with sections, the last section is copied.
- `reduction(op:variable)`: Each thread has its own copy of the specified variable. Reductions can appear only in reduction operation. All copies are "reduced" back into the original master thread's variable.

By default, in do/for and parallel do/for constructs, the index variable is automatically private. Variables scoped within the construct (e.g., automatic variables in functions) are private by default. In the following examples, the index for the for/do loops is private and "a" is shared. All threads have access to the same storage areas for "a," but each memory location is updated by a single thread.

Fortran	C/C++

`!$omp parallel do shared(a), &`
`        private(t1,t2)`
`   do i = 1,1000`
`      t1 = f(i); t2 = g(i)`
`      a(i) = sqrt(t1**2 + t2**2)`
`   end do`
	`#pragma omp parallel for \`
	`   shared(a), private(t1,t2)`
	`for(i=0; i<1000; i++){`
	`   t1 = f[i];  t2 = g[i];`
	`   a[i] = sqrt( (t1*t1 + t2*t2);`
	`}`

An operation that "combines" multiple elements to form a single result, such as a summation, is called a reduction operation. A variable that accumulates the result is called a reduction variable. In parallel loops reduction, operators and variables must be declared. Consider the following code snippet:

Fortran	C/C++

`real*8 asum, aprod`
`     ...`
`asum=0.0; aprod=1.0;`
`do i=1,n`
`   asum  = asum  + a(i)`
`   aprod = aprod * a(i)`
`enddo`
`print*, asum,  aprod`
	`double asum,  aprod`
	`     ...`
	`asum=0.0; aprod=1.0;`
	`for(i=0; i<n; i++){`
	`   asum  = asum  + a[i];`
	`   aprod = aprod * a[i];`
	`}`
	`printf("%f %f\n", asum,aprod);`

Both asum and aprod are updated in each iteration. To correctly parallelize with OpenMP, reduction operations need to be performed on asum and aprod. Otherwise, running with different numbers of threads can potentially produce different results due to race conditions (more on this later in this chapter).

Fortran	C/C++

`asum=0.0; aprod=1.0;`
`!$omp  parallel`
`   ...`
`!$omp do reduction(+:asum ) &`
`!$omp    reduction(*:aprod)`
`     do i=1,n`
`        asum  = asum  + a(i)`
`        aprod = aprod * a(i)`
`     enddo`
`   ...`
`!$omp end parallel`
`print*, asum,  aprod`
	`asum=0.0; aprod=1.0;`
	`#pragma omp parallel`
	`{    ...`
	`#pragma omp for \ reduction(+:asum) \`
	` reduction(*:aprod)`
	`    for(i=0; i<n; i++){`
	`       asum  = asum  + a[i];`
	`       aprod = aprod * a[i];`
	`    }`
	`...`
	`}`
	`printf("%f %f\n", asum,aprod);`

With the reduction clause, each thread has a private copy of asum and aprod, initialized to the operator's identity, 0.0 and 1.0, respectively. After the loop execution, the master thread collects the private values of each thread and finishes the (global) reduction with a summation operation for asum and a product for aprod.

The following example illustrates reductions, lastprivate, and firstprivate. In the first loop, each thread's copy of sum is added to the original sum at the end of the loop. In the second loop, the last iteration value (not the value from the last thread to finish) of tmp is copied back to master; thus, tmp will be 100. In the last loop, each thread will repeat (i.e., pick up) work until the work function returns false.

Fortran

```
sum = 10.0
!$omp parallel do &
!$omp reduction(+:sum)
      do i = 1, 10
         sum = sum + a(i)
      end do
!$omp parallel do &
!$omp lastprivate(tmp)
      do i = 1, 100
         tmp = a(i)
      end do
      print *, 'a(100) = = ', tmp

logical :: torf=.true.
!$omp parallel firstprivate(torf)
  do while(torf)
      torf  = do_work()
end do
!$omp end parallel
```

C/C++

```
sum = 10.0;
#pragma omp parallel for \ reduction(+:sum)
      for (i=0; i<10; i++)
        sum = sum + a[i];
#pragma omp parallel for \ lastprivate(tmp)
      for(i=0; i<=100; i++){
          tmp = a[i];
       }
       print *, 'a(100) = = ', tmp

int torf = 1;
#pragma omp parallel \ firstprivate(torf)
{
  while(torf)
     torf  = do_work();
}
```

13.7.1 Synchronization and Mutual Exclusion

To synchronize execution across threads, OpenMP provides three explicit constructs. These are critical, atomic, and barrier.

With a block of code marked as critical, all threads execute the block of code, but only one thread can execute the block at any given time. In other words, if a thread is currently executing inside a critical region and another thread reaches that critical region and attempts to execute it, it will block until the first thread exits that critical region. The optional name enables multiple different critical regions to exist. Names act as global identifiers: different critical regions with the same name are treated as the same region. All critical sections that are unnamed are treated as the same section.

Atomic directives are similar to critical directives but apply only to a single, simple assignment statement that updates a scalar variable. Primarily designed to be implemented with machine instructions that perform "read, modify, and write" operations on memory atomically, atomic directives are cheaper to execute than critical directives.

Critical directive	Atomic directive
!$omp parallel shared(sum,x,y) ... **!$omp critical** call update(x) call update(y) sum=sum+1 **!$omp end critical** ... **!$omp end parallel**	**!$omp parallel** ... **!$omp atomic** sum=sum+1 ... **!$omp end parallel**

Note that atomic applies to the "sum=sum+1" operation only and cannot be used with a more complex statement.

OpenMP barriers can be placed within parallel regions to synchronize threads. No thread will proceed past the barrier until all threads arrive at the barrier. Consider the following example:

```
#pragma omp parallel shared (A, B, C) private(id)
{
   id=omp_get_thread_num();
   A[id] = big_calc1(id);

   #pragma omp barrier   // explicit barrier
   #pragma omp for
   for(i=0;i<N;i++){
           C[i]=big_calc3(i,A,N);
   } // implicit barrier
   #pragma omp for nowait
   for(i=0;i<N;i++){
           B[i]=big_calc2(C, I, N);
   }
   A[id] = big_calc4(id);
}
```

The function `omp _ get _ thread _ num` returns the integer id number of the thread. In this case, each thread will call `big _ calc1`, store the result in `A[id]`, where `id` is the thread id, and wait at the barrier for all the other threads to catch up. Once all the threads arrive at the barrier, threads will start executing iterations from the first for loop. Threads will again wait at the implicit barrier at the end of the first for loop before proceeding to the final for loop. At the final for loop, the threads have a `nowait` statement that bypasses the implicit barrier at the end of the for loop. Some threads might be executing `big _ calc4`, while other threads are performing iterations in `big _ calc2`.

13.8 More Language Constructs

13.8.1 Runtime Library Application Programming Interface

The OpenMP runtime library also includes a few functions that aid in parallelization, report application state, and even set runtime controls. To use these libraries, the header file "omp.h" should be included in C/C++ and "omp_lib.h" should be included in Fortran. In Fortran 90, the module "omp_lib" should be used.

C/C++ use include file	#include <omp.h>
Fortran use include file	include "omp_lib.h"
Fortran 90 use the module file	use omp_lib

The following functions are available to the user in the runtime library.

Function	Operation
omp_get_num_threads()	Number of threads in team, returns an integer (N)
omp_get_thread_num()	Thread ID. {0 -> N-1}
omp_get_num_procs()	Number of machine cores
omp_in_parallel()	True if in parallel region and multiple thread executing
omp_set_num_threads(#)	Changes number of threads for parallel region
omp_get_dynamic()	True if dynamic threading is on
omp_set_dynamic(int)	Set state of dynamic threading (1/0 for true/false)
omp_get_wtime()	Returns a double/real with the elapsed wall clock time in seconds

The omp_get_wtime() function is of particular importance in timing OpenMP code and as a matter of good practice should be included and used to time all OpenMP applications. An example of its use would be as follows:

```
#include<stdio.h>
#include<omp.h>
int main(int argc, char ** argv)
{
  double tick, tock;
  tick = omp_get_wtime();
  #pragma omp parallel
  {
    /* perform some work here */
  }
  tock = omp_get_wtime();
  printf("The time spent in OpenMP parallel region was: %f\n",
tock-tick);
  return 0;
}
```

13.8.2 Environment Variables

We have already encountered the OMP_NUM_THREADS environment variable, which controls the number of threads created in parallel regions. Other environment variables are available and can be used to change the behavior of OpenMP constructs within the application without having to recompile the source code.

It is worth mentioning that environment variables typically have the lowest priority of all OpenMP constructs. For example, if the code internally sets the number of threads with omp_set_num_threads(), changing the number of threads with the OMP_NUM_THREADS environment variable will not change the number of threads that the application is run with.

The following environment variables are available in most OpenMP environments:

OMP_DYNAMIC	Specifies whether the OpenMP runtime library can adjust the number of threads in a parallel region.
OMP_NESTED	If enabled, nested for and do loops are distributed across threads.
OMP_NUM_THREADS	Sets the maximum number of threads in a parallel region.
OMP_SCHEDULE	Modifies the default scheduling clause for all schedule (runtime) workshares.

A typical use case for OMP_SCHEDULE would be to benchmark an OpenMP application with omp_get_wtime() and different scheduling clauses: static, dynamic, and guided. This would remove the need to recompile the source code for every experiment.

```
#include<stdio.h>
#include<omp.h>
int main(int argc, char ** argv)
{
   double tick, tock;
  tick = omp_get_wtime();
  #pragma omp parallel
  {
    #pragma omp for schedule(runtime)
      for (int i = 0; i < 1000; ++i) // some big loop
        {
          /* do some work */
      }
  }
  tock = omp_get_wtime();
  printf("The time difference was: %f\n", tock-tick);
  return 0;
}
```

There are other environments that are available to specific compilers (GNU and Intel compilers) and specific machines (Cray systems). Some are for more advanced users such as OMP_STACK_SIZE, which control the size of the stack for created (i.e., not master) threads, and some are used purely for debugging purposes (GOMP_DEBUG). For these advanced and compiler/system-specific environment variables, good programming practices ought to be followed: keep it simple, and apply judiciously.

13.9 Practical Issues

13.9.1 Race Conditions

A race condition is a subtle error where multiple threads may potentially update the same memory location, causing a wrong result. The application will run successfully but on occasion return the wrong results. This is a very dangerous error to encounter and can easily go undetected. The following tests can help expose and diagnose this sort of error:

- Running the OpenMP application repeatedly produces different results.
- Running the application with different numbers of threads produces different results, specifically from the single thread (`OMP _ NUM _ THREADS=1`) to any number of threads.
- Changes to the scheduling produce different results.
- Using the `ordered` clause in a `do/for` workshare construct produces the correct result, but removing it produces different results.
- Adding a call to sleep (to simulate processor load or increased workload) in one OpenMP construct affects the result.

A simple example of a race condition would be as follows:

Fortran	C/C++

`program main`
`implicit none`
`integer count`
`count=0`
`!$omp parallel`
`count=count+1`
`!$omp end parallel`
`write (*,*) "count is: ", count`
`end program main`
	`#include<stdio.h>`
	`int main(int argc, char **argv)`
	`{`
	`  int count=0;`
	`  #pragma omp parallel`
	`  { count=count+1;}`
	`  printf("count is: %d\n",count);`
	`}`

Since `count` is shared across threads and appears in both the left- and right-hand sides of an assignment operation, one thread might be reading it while another is updating it, leading to an inconsistent value. The same effect can appear in C/C++ with an increment operator: `count++` or `++ count`. To fix this race condition, a single or master clause needs to be added around count so that the operation happens only once to yield the same result as running the application sequentially without OpenMP.

Similarly, consider the following code snippet, where `a[]`, `b[]`, `c[]`, and `d[]` are all `N-sized` arrays initialized to 1.

```
#pragma omp parallel \ shared(a,b,c,d) private(i)
{
  #pragma omp for nowait
  for(i=0; i<N;i++) a[i]  = b[i]+c[i];
  #pragma omp for
  for(i=0; i<N;i++) d[i]  = a[i]*b[i];
}
```

A single thread running either code will set a[] to 2 (1+1) and d[] to 2 (2*1). Running in parallel will yield different values for some of d[]. This is because a[] is shared across OpenMP constructs and appears in both the left- and right-hand sides of assignment operations and two or more threads are computing to update-and-use its value.

As a best practice guideline, it is critical to understand the data scoping of every variable that appears in an OpenMP construct. Manually assigning the data scoping rules, what is private, what is shared, and what is reduced, helps enforce thinking about race conditions and the pitfalls of parallelism in code execution, especially for variables such as a[], which appear on both the left- and right-hand sides of assignment operators.

To solve the race condition, we should impose thread synchronization either through a barrier (implicit or explicit), a critical/atomic, or an OpenMP lock. The simplest solution would be to restore the implicit barrier by removing the nowait clause.

This is not to say that there are no tools available for syntactic and dynamic code analysis to detect possible race conditions since some tools, with varying capabilities, do exist. However, none of them can guarantee full race condition detection. There simply is no substitute to fully understanding code behavior and adequately frequent testing.

13.9.2 Reducing Errors

Code development is an error-prone process, and multithreaded code is even more so due to the complexity of parallelism. The following is an incomplete list of common beginner errors.

- Missing compiler option (forgot `-openmp` or `-fopenmp`).
  - The program executes sequentially, therefore slower than expected.
  - Manifests in 0 valued time reports from `omp_get_wtime()`.
- Missing `parallel` keyword, no threads created, therefore program executes sequentially.
- Missing `omp` keyword: Compiler directives are comments, therefore no parallelization.
- Missing `do/for` keyword: Loops are not distributed across threads, poor performance.
- Missing/forgotten `reduction`: Results depend on the number of threads used.
- Missing/forgotten `private` clause or incorrect variables.
- Misuse of the `lastprivate/firstprivate` clause.
- Unnecessary or unintentional parallelization: This can reduce performance for small loops (use conditions) or cause race conditions.
- Concurrent use of a `shared` variable on both sides of assignment: This is a race condition.
- Uninitialized local variables in a dynamic scope.
- Parallel array processing without iteration ordering when the order of iterations is important: Adding/removing the `ordered` clause will highlight this issue.

There are several high-quality commercial and open source tools available that can drastically reduce debugging time, should you encounter any problems. These include the following:

- Allinea Distributed Debugging Tool (DDT) debugger, a parallel commercial debugger
- Intel thread checker: To catch race conditions
- Valgrind [5]: To catch race conditions
- Data Display Debugger (DDD) [6]: A thread capable interface to gdb
- gdb: A ubiquitous debugger available on all Unix-like systems

Not all tools are available on all platforms, but gdb and Valgrind are fairly common, with plenty of examples and online tutorials which are beyond the scope of this introduction. These tools are not a substitute to fully understanding the intended code behavior, having tests with adequate coverage, and of course understanding every OpenMP added construct.

To get the most out of OpenMP, please take note of the following:

- Merge parallel sections to avoid the cost of thread creation/destruction.
- Use atomic instead of critical due to its lower cost.
- Don't overprotect local or private variables; these are private to every thread.
- Don't overprotect a code fragment executed by a single thread.
- Avoid having too much code in a critical section:
    - If it doesn't use shared variables, it shouldn't be in a critical section.
    - Avoid complex function calls, keep it simple.
- Avoid having too many entries to a critical section.
    - If you have a critical section that contains an if-statement, move the if-statement to outside the critical section.

13.10 Concluding Remarks

There are distinct advantages to OpenMP, most notably the fact that an application can have OpenMP constructs added incrementally without major architectural changes. It is also easy to use, making it ideal for adding multithreading capability to an application in a short development cycle. OpenMP is, however, a shared memory platform, and it cannot scale beyond a single workstation/server/compute-node. To scale past that limit, developers will have to use the Message Passing Interface (MPI) library.

References

1. OpenMP specification: http://openmp.org/wp/openmp-specifications/.
2. TBB: https://www.threadingbuildingblocks.org/.
3. CilkPlus: https://www.cilkplus.org/.

4. GNU compilers: https://gcc.gnu.org/.
5. Valgrind: http://valgrind.org/.
6. DDD debugger: https://www.gnu.org/software/ddd/.

Further Reading

Victor Eijkhout's *Introduction to High-Performance Scientific Computing*: http://pages.tacc.utexas.edu/~eijkhout/istc/istc.html.

14

Checkpointing Code for Restartability with HDF5

Frank T. Willmore

CONTENTS

14.1 Checkpointing and Restarting

Checkpointing is a technique for providing fault-tolerance to a code, particularly when the code runs on a complex system with multiple potential points of failure, and which consumes considerable time and/or resources. It involves taking a snapshot of the state of the program and the values of all variables that will be accessed as the program continues to run and saving them, typically to disk, so that if a failure (power outage, hardware fault, network outage, etc.) occurs, the program may be restarted from this saved state information. In the move toward ever more parallelism, the issues of fault tolerance become more poignant. The larger and more complex a computing system is, and the more processors, circuits, and others, the greater the possibility of failure at some point in the system during the course of simulation.

Major research computing facilities are typically one-of-a-kind installations and experience failures as a regular part of operations. As such, they typically now require that user code employ an appropriate level of checkpointing and will refuse to reimburse beyond a certain number of hours of execution, even if the failure is ostensibly due to the service provider. If a code runs for 10 minutes on a laptop, it probably doesn't need the checkpointing and restart capabilities. However, if it runs for 100 hours on 1024 or more processors on a research cluster, it most certainly does.

A typical checkpointing scheme might have the following requirements:

- Ability to save the state of the simulation at regular intervals
- Ability to restart the simulation, given the executable and a restart file
- Ability to read simulation data and possibly restart on a different platform
- Ability to examine contents of a restart file
- Ease of implementation

14.2 HDF5

Hierarchical Data Format in its fifth incarnation (HDF5) is an emergent standard for the storage, retrieval, and sharing of data, scientific and otherwise, in much the same spirit as PDF is a standard for digitization of a printable document, an RDBMS is for the storage and retrieval of relational data, XML is a standard for a readable markup of document encoding, tar-zip is for a file archive, and so on. In addition to data, a file in HDF5 also stores metadata, and one file can hold as many datasets as necessary. The primary purpose of HDF5 is to abstract the hardware layer away from the user. Instead of worrying about endianness, or how best to read/write data in parallel, or on whatever specific hardware you are running, you are now free to work on your science, financial analysis, and others, and let the good folks at the HDF Group* do the heavy lifting when it comes to the hardware. By writing data to an HDF formatted file, you can be reasonably assured that your collaborators, rival researchers, program officer, or perhaps some future version of yourself will be able to read the HDF data file you wrote, regardless of what hardware they use, what operating system they run, or in what language they write code. The example in this chapter uses Python because the h5py Python to HDF5 interface is very easy to use and illustrates the capabilities. Application programming interface (API) exist for Python, C, C++, Fortran, and Java, and others are being developed. Also, HDF5 is developed with future compatibility in mind, so a file written in HDF5 today and read 10–20 years later has a reasonable expectation of still being fully legible and compatible.

Although the calls in this example are written in Python, the underlying library doing the complicated memory access is written in C, and the performance is inherently optimized at the C language level, which is very close to the hardware. Considering that support for parallel I/O is also handled at this same level, this is a lot of I/O optimization you are leveraging by using HDF5.

14.3 H Is for Hierarchical

Flat files are a great tool and, in combination with standard posix utilities, can do a great deal of work. But at some point, you will need to increase your level of sophistication, and HDF5 allows you to store numbers, images, sound recordings, video, simulation results, and literally any other digital information in a flexible and convenient format, in a single, modifiable file, in parallel, and gives the best promise of allowing such information to be read at a later date.

HDF uses a hierarchical POSIX path syntax. It not only looks like a filesystem on the inside ("/path/to/data") but it is in fact possible to store files as well as other digital objects (photos, audio, etc.) in that tree and to attach *attributes*, or metadata, that describe important information about the data, such as how it was collected, what the temperature was, and anything that you deem important.

HDF5 is an important file format and is gaining increasing traction in the research community for a few very important reasons. With the rise of simulation and digital data-based science, it becomes overwhelmingly important to the peer review process to be able

* www.hdfgroup.org.

to reproduce digital results, as well as to share data. The introduction of digital standards lays the groundwork for the successful sharing of data, while eliminating much miscommunication regarding data types, precision standards, data ranges, ordering, formats, and a myriad of other issues that crop up in digital work. HDF5 has emerged as a de facto standard in part because it provides mechanisms to handle these issues.

14.4 Installing and Using HDF5

Installing HDF5 is system specific, and if you are running at a large service provider, it is likely that they will already have an installed and supported version. If you maintain your own system, you will need to install either the parallel or serial version (depending on your setup and needs) or the Python interface; for example, on a Debian Linux system, this should be as simple as

```
$ sudo apt-get install libhdf5-serial-dev
$ sudo apt-get install hdf5-tools
$ sudo apt-get install python3-h5py
```

This will install the underlying HDF5 library, the tools, and the h5py Python interface. For simplicity, our example here is with Python and uses the h5py interface. It is worth mentioning that when working directly with the C or C++ language interfaces, HDF5 includes some helpful scripts (h5cc, h5fc, and h5cpp) that wrap the compiler so that the correct libraries, headers, and others, are included. This can also be done by hand, but these scripts are handy and work analogously to mpi wrapper scripts mpicc, mpif90, and others, with which you may already be familiar.

14.5 A Simulation Example Using HDF5 for Checkpointing and Restarting

Take the following simple one-dimensional (1-D) diffusion simulation. We start with an approximate Dirac-delta distribution of 10,000 atoms/unit at a single point roughly at the middle (position 512) of a simple 1-D lattice with 1024 total positions. At every time step, each atom can either stay still or hop away, based on a simulated coin toss. If it hops, it goes either to the left or to the right, also based on a simulated coin toss. We will checkpoint (save the state of) the system at every time step to an HDF5 file. We will then restart the program from a given time step by specifying a restart file and time step to the simulation code.

The code is in Python and imports the h5py Python to the HDF5 interface, the numpy numerical package, argparse for parsing command line options, and the platform that grants access to system info, which will be recorded as metadata in the restart file. The entire code is contained in the file simulation.py, shown here:

```
1 #!/usr/bin/env python
2
3 import h5py
4 import numpy as np
```

```
5 import argparse
6 import platform
7
```

There is a function that opens an HDF5 formatted restart file, writes the current simulation state to that file, and then closes the file. The file mode 'a' means that additional data can be added to an existing file, and that if the file does not already exist, it gets created:

```
 8 def write_frame_to_restart_file(restart_file_name, time_step,
 9                                  lattice):
10     restart_file = h5py.File(restart_file_name, 'a')
11     restart_file["/" + simulation_name + "/" + str(time_step)]
12                       = lattice
13     restart_file.close()
14
```

Default values are provided for the number of steps to run the simulation: the restart_time, which tells the system at which point in its history the simulation should be restarted; the restart_file_name, which tells which file should be used to restart the simulation; and finally the simulation_name, since a given restart file can store the results of more than one simulation:

```
15 # default values
16 run = 10
17 restart_time = 0
18 restart_file_name = "restart.h5"
19 simulation_name = "foo"
20
```

Command line options are parsed, and default values are overwritten with passed values:

```
21 # Parse command line arguments
22 parser = argparse.ArgumentParser()
23 parser.add_argument('--run', nargs=1, type=int,
24                     help='run help')
25 parser.add_argument('--restart_file_name', nargs=1,
26                     help='name of restart file')
27 parser.add_argument('--restart_time', nargs=1, type=int,
28                     help='restart the simulation at ...')
29 parser.add_argument('--simulation_name', nargs=1,
30                     help='simulation name')
31 args = parser.parse_args()
32 if args.run != None:
33     run = args.run[0]
34 if args.restart_time != None:
35     restart_time = args.restart_time[0]
36 if args.restart_file_name != None:
37     restart_file_name = args.restart_file_name[0]
38 if args.simulation_name != None:
39     simulation_name = args.simulation_name[0]
40
```

The restart file is opened and a node is added for the current simulation, if it does not already exist. Metadata (in this case, information about the system and release of the operating system of the platform hosting the simulation) is recorded in the HDF5 file as an *attribute* associated with the simulation name:

```
41 restart_file = h5py.File(restart_file_name, 'a')
42
43 # create node if it doesn't exist
44 if not "/" + simulation_name in restart_file:
45     restart_file.create_group("/" + simulation_name)
46 # Write metadata to restart file as attributes
47 restart_file["/" + simulation_name].attrs.create(
48     "system", platform.system())
49 restart_file["/" + simulation_name].attrs.create(
50     "release", platform.release())
51
```

Two scenarios are possible. Either the code is being invoked as a restart, in which case the state is loaded here from the HDF5 restart file, or it is invoked as a new simulation, in which case the initial condition is specified explicitly and then recorded as the first frame for the given simulation name in the restart file.

```
52 if args.restart_time != None: # load state from restart file
53     restart_time = args.restart_time[0]
54     lattice=restart_file["/" + simulation_name + "/"
55           + str(restart_time)]
56     time_step = restart_time
57
58 else: # generate new simulation
59     lattice=np.zeros(1024, dtype=np.int)
60     lattice[512] = 10000
61     time_step = 0
62     # delete if it exists
63     if "/" + simulation_name + "/0" in restart_file:
64     restart_file.__delitem__("/" + simulation_name + "/0")
65     write_frame_to_restart_file(restart_file_name, time_step,
66                                 lattice)
67
```

For this example, it is possible to restart a simulation at any point for which data exist. For example, even though the simulation may have run for 20 steps, it can still be restarted from step 15. Any steps beyond the specified restart step are deleted here.

```
68 # delete any forward steps from the restart file
69 restart_time += 1 # begin deleting records from here
70 next_dataset = "/" + simulation_name + "/" + str(restart_time)
71 while next_dataset in restart_file:
72     restart_file.__delitem__(next_dataset)
73     restart_time += 1
74     next_dataset = ("/" + simulation_name + "/"
75                    + str(restart_time))
76
```

In this example, the actual code of the simulation (shown in the following) is quite small, and the entire state of the simulation is contained in the variable lattice[]. These values become recorded in the restart file:

```
77 # run the simulation
78 while time_step < run:
79     new_lattice=np.zeros(1024, dtype=np.int)
80     for i in range(1024):
81             for j in range(lattice[i]):
82              sample = np.random.random()
83              if ( i > 0 and i < 1023):
84                      if (sample <= 0.25): new_lattice[i - 1] += 1
85                      elif (sample >= 0.75): new_lattice[i + 1] += 1
86                      else: new_lattice[i] += 1
87     # save for this time step
88     lattice = new_lattice
89     time_step = time_step + 1
90     write_frame_to_restart_file(restart_file_name, time_step,
91                                     lattice)
92 # end of file
```

14.6 The HDF5 Restart File

In order to look at the HDF5 restart file, run the simulation using the default values for simulation name and restart file name and number of steps:

```
$ ./simulation.py
$
```

The code itself generates no output at the command line, but the HDF5 restart file contains all the data. We can see it by using the command line utilities h5ls, which gives an 'ls'-like listing of the file contents, and h5dump, which simply dumps all of the info contained in the file to the screen. Running h5ls at the top level of the restart file, we see that it contains one group, foo:

```
$ h5ls restart.h5
foo                    Group
```

Going deeper, we see the group `foo` contains datasets for each of the time steps of the simulation thus far:

```
$ h5ls restart.h5/foo
0               Dataset {1024}
1               Dataset {1024}
10              Dataset {1024}
2               Dataset {1024}
3               Dataset {1024}
4               Dataset {1024}
5               Dataset {1024}
```

```
6               Dataset {1024}
7               Dataset {1024}
8               Dataset {1024}
9               Dataset {1024}
```

Using h5dump, we see that the group "foo" contains the metadata attributes "release" and "system" that we assigned to it. We also see metadata about the metadata... various info on the data type used to hold the values we assigned, in this case release = "3.13.0-32-generic" and system = "Linux." We also see the first dataset "/foo/0" and the values it contains, truncated here for brevity.

```
$ h5dump restart.h5 |head -n 40
HDF5 "restart.h5" {
GROUP "/" {
   GROUP "foo" {
       ATTRIBUTE "release" {
       DATATYPE  H5T_STRING {
              STRSIZE 17;
              STRPAD H5T_STR_NULLPAD;
              CSET H5T_CSET_ASCII;
              CTYPE H5T_C_S1;
       }
       DATASPACE  SCALAR
       DATA {
       (0): "3.13.0-32-generic"
       }
       }
       ATTRIBUTE "system" {
       DATATYPE  H5T_STRING {
              STRSIZE 5;
              STRPAD H5T_STR_NULLPAD;
              CSET H5T_CSET_ASCII;
              CTYPE H5T_C_S1;
       }
       DATASPACE  SCALAR
       DATA {
       (0): "Linux"
       }
       }
       DATASET "0" {
       DATATYPE  H5T_STD_I32LE
       DATASPACE  SIMPLE { ( 1024 ) / ( 1024 ) }
       DATA {
       (0): 0, 0, 0, 0, 0, 0, 0, 0, 0, 0, 0, 0, 0, 0, 0, 0, 0, 0, 0, 0, 0,
       (21): 0, 0, 0, 0, 0, 0, 0, 0, 0, 0, 0, 0, 0, 0, 0, 0, 0, 0, 0, 0, 0,

       etcetera...
```

At first glance, this may look like a lot of extra stuff that gets included in to the restart file. And arguably it is. The point to take away is that you can still use it to do simple tasks, like write a restart file. Just the fact that you use HDF5 to record the info means that a lot of metadata get recorded and that you don't have to worry about whether the size of the type and everything else is being done correctly; it is taken care of for you.

14.7 Restart

So, now we see what gets recorded. Let's try the restart, and for kicks, let's now run it under a different name:

```
$ ./simulation.py --simulation_name bar --run 5
```

No output, but hopefully the restart info will get written. Let's look:

```
$ h5ls restart.h5
foo              Group
bar              Group
```

So far, so good. There is now a new entry for the newly named 'bar' simulation. Examining the bar group, we see datasets recorded for each time step:

```
$ h5ls restart.h5/bar
0                Dataset {1024}
1                Dataset {1024}
2                Dataset {1024}
3                Dataset {1024}
4                Dataset {1024}
5                Dataset {1024}
```

Now, we run the restart and look at the bar group again and see that it contains info for all the new time steps:

```
$ ./simulation.py --simulation_name bar --restart_time 5 --run 10
$ h5ls restart.h5/bar
0                Dataset {1024}
1                Dataset {1024}
10               Dataset {1024}
2                Dataset {1024}
3                Dataset {1024}
4                Dataset {1024}
5                Dataset {1024}
6                Dataset {1024}
7                Dataset {1024}
8                Dataset {1024}
9                Dataset {1024}
```

Restart capability can be created with a minimal amount of additional coding. And using the flexibility and portability of HDF5 guarantees that the file will be accessible across platforms, contain appropriate metadata, and be readable for many generations to come. In short, HDF5 is highly configurable and flexible.

14.8 A Note on Stochastic Methods and Restart

The reproducibility of a result for a stochastic simulation depends on the reproducibility of the results of the random number generator and, in the case of parallel computing, may depend on the order in which the elements of a reduction operation are combined. In this example, the state of the random number generator was not saved. This means that in this simulation, the results will be statistically equivalent but not identical. This is relevant to and best addressed under the broader topics of stochastic simulation, but it bears mentioning here. If you like, take it as an exercise and add, as a feature, the ability to specify a new seed value, and then create a new restart file for a new time step.

15

Libraries for Linear Algebra

Victor Eijkhout

CONTENTS

15.1 Introduction

Linear algebra is a fundamental building block of nearly every scientific software package: Adding vectors, calculating eigenvalues, and multiplying matrices underlie everything from molecular dynamics, to fluid mechanics, to finite element (FE) simulations. Due to the ubiquity of linear algebra in scientific computing, many new computational scientists may be tempted to write their own dot product routine, for example, when necessitated by their project. The purpose of this chapter is to show how to resist this temptation by providing an introduction to fast, free libraries for doing linear algebra that will save time and frustration. We discuss the Basic Linear Algebra Subprograms (BLAS), Linear Algebra Package (LAPACK), and Portable Extendable Toolkit for Scientific Computing (PETSc) linear algebra libraries, each of which provides capabilities for solving linear algebra problems of different sizes and structure.

Two reasons for using libraries instead of coding an algorithm from scratch are programmer productivity and code quality. Once an algorithm has been shown to be correct and included in a library, it permits programmers to focus on *using* the algorithm to do useful science, rather than debugging code that already exists elsewhere. Another reason to use libraries is for their performance: The optimized algorithms provided by specialized libraries can be orders of magnitude faster than naive implementations. In fact, because BLAS and LAPACK [1] are now *de facto* standards for linear algebra, hardware vendors (e.g., Intel and AMD) create optimized versions of BLAS and LAPACK libraries that are specifically tuned for their central processing unit (CPU) instruction sets. From a user perspective, this allows you to write scientific software in an architecture-agnostic way and easily port your code to other CPU platforms by compiling your code against the appropriate BLAS or LAPACK library.

The three problem scales and corresponding packages discussed in this chapter are as follows:

1. The Basic Linear Algebra Subprograms (BLAS) library contains the basic operations between vectors and matrices; these are conceptually the simplest but also the most performance-critical.
2. The Linear Algebra Package (LAPACK) then builds on the BLAS to offer linear system solving and eigenvalue routines. Parallelism in LAPACK is limited to multicore parallelism.
3. Large-scale, distributed memory, parallelism introduces another layer of complexity [2,3]. This is covered by libraries such as PETSc.

15.2 Choosing a Library

Because there are many implementations of linear algebra libraries for different processor architectures and development environments, choosing a library can feel a bit daunting. A reference implementation of BLAS and LAPACK that is easy to download and install is available on the Netlib website (http://www.netlib.org). The Netlib implementation is useful for learning how to use BLAS and LAPACK but is not optimized for performance. Similarly, the GNU Scientific Library (GSL: https://www.gnu.org/software/gsl/) provides a nonoptimized BLAS implementation that is easy to install via most package managers and can easily be linked against an optimized library like ATLAS. Popular implementations tuned for performance include the following, among many others:

- The Intel Math Kernel Library (MKL), which comes with the Intel compiler: https://software.intel.com/en-us/intel-mkl
- The AMD Core Math Library: http://developer.amd.com/tools-and-sdks/archive/amd-core-math-library-acml/
- The BLAS-like Library Instantiation Software (BLIS) library contains micro kernels for many popular architectures: https://github.com/flame/blis

- The Automatically Tuned Linear Algebra Software (ATLAS) library provides C and Fortran Application Programming Interfaces (APIs) for BLAS, but only a subset of the LAPACK API: http://math-atlas.sourceforge.net/
- vecLib, the Apple implementation of BLAS/LAPACK: https://developer.apple.com /library/mac/documentation/Performance/Conceptual/vecLib/
- The cuBLAS library for NVIDIA graphics processing units (GPUs): https://developer .nvidia.com/cuBLAS
- The clBLAS libraries for heterogeneous computing platforms: https://github.com /clMathLibraries/clBLAS

For the new user, choose an implementation that is easy to install in your development environment. The key takeaway is that using *any* BLAS/LAPACK implementation is better than none, and after developing code that uses these premade routines, you can always try building against a different library to compare performance.

15.3 BLAS

Just as there are three levels of hierarchy for problems solved by BLAS, LAPACK, and PETSc, the BLAS libraries themselves are decomposed into problems at three levels:

1. BLAS Level 1 operations work on vectors and have a typical complexity of n operations for a vector of length n. Typical examples are vector addition, computing an inner product, or applying Givens rotations [4].
2. BLAS Level 2 operations are of the matrix–vector type, with n^2 complexity, such as the matrix–vector product.
3. Finally, BLAS Level 3 operations are between two matrices, the typical example being the matrix–matrix product, which has $O(n^3)$ complexity.

There are hundreds of different BLAS routines, some of which are closely related. In order to help differentiate the different routines while keeping the names of the routines relatively concise, BLAS employs a structured naming scheme of the form:

$$\texttt{XYYZZZ:}\begin{cases}\texttt{X : precision}\\ \texttt{YY : matrix type}\\ \texttt{ZZZ : operation}\end{cases}$$

Here, the "precision" character can be `S` for single, `D` for double, `C` for single-precision complex, and `Z` for double-precision complex. The two characters indicating matrix type can describe both storage and mathematical properties. For instance, `GE` stands for general rectangular and `GT` for general triangular, while `PB` stands for a positive definite band matrix and `PO` for positive definite. The (up to three) characters used to describe the operation are abbreviated with shorthands, including `MM` for matrix multiply or `TRF` for triangular factorization. Knowing these rules helps with finding information about routines you wish to use in your code. For example, if you know you want to multiply two double-precision, general rectangular matrices, you can apply the rules above to generate `dgemm` as a possible routine to use and then search for "dgemm documentation."

The BLAS naming convention is an artifact of early versions of Fortran, where function names were limited to six characters. The original BLAS routines were written in Fortran, and the reference implementation is still written in Fortran today. Calling BLAS routines is therefore straightforward in Fortran. To use the Fortran routines from a C/C++ program, most users will utilize an interface to the Fortran routines provided by a library or by calling the Fortran functions directly (and doing the appropriate function-name and argument modifications, as well as appropriate linking).

15.3.1 Vector Operations

To demonstrate how BLAS routines can be used in your programs, we begin with a code snippet for scaling the components of a vector. In this example, we consider a double-precision vector `xarray` and will use `dscal` to multiply each element of this vector by the same value:

```
// example1.f90
do i=1,n
  xarray(i) = 1.d0
end do
call dscal(n,scale,xarray,1)
do i=1,n
  if (.not.assert_equal( xarray(i),scale ))
    print *,"Error in index",i
end do
```

The `dscal` function arguments are the length of the vector (`integer`), the factor to scale by (`double precision`), the vector to scale (`double precision, dimension(*)`), and stride (`integer`), respectively. The stride specifies the size of the increment step. For example, if the stride was 2 instead of 1 as in the above example, only the odd elements of the vector would be scaled. A snippet from a C++ implementation of the above Fortran90 example might look like the following:

```
// example1c.cxx
xarray = new double[n]; yarray = new double[n];
for (int i=0; i<n; i++)
  xarray[i] = 1.;
cblas_dscal(n,scale,xarray,1);
for (int i=0; i<n; i++)
  if (!assert_equal( xarray[i],scale )) &
    printf("Error in index %d",i);
```

15.3.2 Matrix–Vector Operations

The BLAS routines for handling matrix operations typically require many more arguments than the functions for manipulating vectors. This is a consequence of matrix elements being stored in consecutive memory locations: the sixth and seventh elements (indices 5 and 6 in Figure 15.1) are in adjacent regions of memory, despite being handled as if they are on separate rows for matrix math. In order for BLAS to handle the shape of a matrix

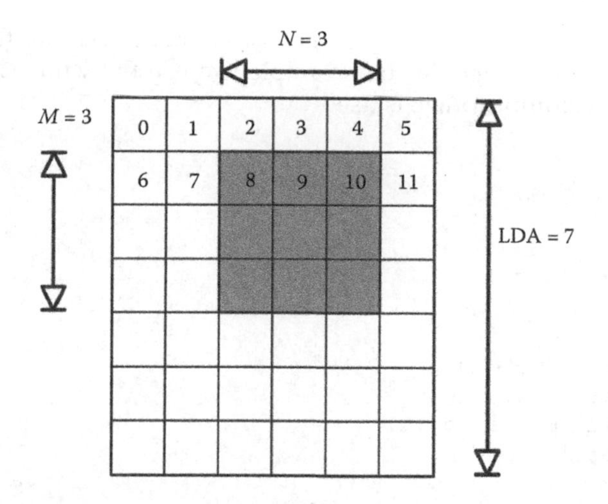

FIGURE 15.1
The elements of a matrix are stored in consecutive elements in memory, so matrix routines need to know the width (N) and height (M) and leading dimension (LDA) to properly handle a matrix's shape. Here, a gray $M = 3$, $N = 3$ submatrix is shaded within a larger $N = 6$, $M = 7$ matrix. The indices of the first 12 elements of the matrix are indicated.

or submatrix (the shaded gray region in Figure 15.1), we must pass the height `M`, width `N`, and the leading dimension `LDA` of the matrix to most matrix routines. Take, for example, `dgemv`, for computing the matrix–vector products:

```
subroutine dgemv(character TRANS,
    integer M,integer N,
    double precision ALPHA,
    double precision, dimension(lda,*) A,integer LDA,
    double precision, dimension(*) X,integer INCX,
    double precision BETA,double precision, dimension(*) Y,integer INCY
    )
```

which computes $y \leftarrow \alpha Ax + \beta y$ for a double-precision matrix A, vector x, vector Y, and scalar α. In addition to the size parameters mentioned above, `dgemv` also requires a character argument `TRANS`, which can be `N` or `Y`, that specifies whether A or its transpose should be used in the calculation. An example of the use of this routine is as follows:

```
// example3.F90
do j=1,n
  xarray(j) = 1.d0
    do i=1,m
      matrix(i,j) = 1.d0
    end do
end do
alpha = 1.d0; beta = 0.d0
call dgemv('N',M,N, alpha,matrix,M, xarray,1, beta,yarray,1)
do i=1,m
  if (.not.assert_equal( yarray(i),dble(n) )) &
    print *,"Error in index",i,":",yarray(i)
end do
```

The same example in C has an extra parameter to indicate whether the matrix is stored in row or column major storage; in this example, we use the value `CblasColMajor` to indicate that Fortran column storage is used:

```
// example3c.cxx
for (int j=0; j<n; j++) {
  xarray[j] = 1.;
  for (int i=0; i<m; i++)
    matrix[ i+j*m ] = 1.;
      }
alpha = 1.; beta = 0.;
  cblas_dgemv(CblasColMajor,
       CblasNoTrans,m,n,
       alpha,matrix,m, xarray,1,
       beta,yarray,1);
for (int i=0; i<m; i++)
  if (!assert_equal( yarray[i],(double)n ))
       printf("Error in index %d",i);
```

To use the `M=3` and `N=3` submatrix at (2, 3) in Figure 15.1 in a matrix routine, we simply pass the top-left element of the submatrix in addition to `M=3`, `N=3`, `LDA=7`. In Fortran code, this would look like the following:

```
double precision :: A(7,6)
integer :: istart=2,jstart=3, M=3,N=3,LDA=7

call SomeBlasRoutine( A(istart,jstart),M,N,LDA ... )
```

15.3.3 Matrix–Matrix Operations

For our next BLAS example, we discuss the multiplication of two matrices using the C-interface to BLAS provided by the GNU scientific library (Listing 15.1). In this case, we will be multiplying two real, double-precision matrices, so `dgemm` is an appropriate function call. The primary difference between the `dgemm` call through the GSL interface and that through Fortran is that the matrix size parameters (in the previous example: `M`, `N`, `LDA`) are not required. On one hand, it is convenient that the `gsl_blas` call to `dgemm` handles the size parameters automatically. On the other hand, it comes at the cost of having to learn how the `gsl_blas` calls to functions like `gsl_matrix_view` and `gsl_bas_dgemm` interface with and vary from the Fortran routines. In short, any conveniences that come from using a particular BLAS interface come at the cost of having to learn the quirks of that particular package, which is a common theme in scientific computing.

LISTING 15.1 `example.c`: A Matrix Multiplication Example Using the GNU Scientific Library BLAS Implementation

```
#include <stdio.h>
#include <gsl/gsl_blas.h>
int main (void){
  double a[] = { 0.1, 0.2, 0.1,
                 0.0, 0.0, 0.1,
                 0.5, 0.1, 0.1};
```

```
    double b[] = { 1.0, -1.0, 0.0,
                   0.0,  0.0, 0.0,
                   2.0,  0.0, 0.0};

    double c[] = { 0.0, 0.0, 0.0,
                   0.0, 0.0, 0.0,
                   0.0, 0.0, 0.0};

    gsl_matrix_view A = gsl_matrix_view_array(a, 3, 3);
    gsl_matrix_view B = gsl_matrix_view_array(b, 3, 3);
    gsl_matrix_view C = gsl_matrix_view_array(c, 3, 3);
    gsl_blas_dgemm (CblasNoTrans, CblasNoTrans,
                    1.0, &A.matrix, &B.matrix,
                    0.0, &C.matrix);
    printf ("[ %f, %f, %f\n", c[0], c[1], c[2]);
    printf ("  %f, %f, %f\n", c[3], c[4], c[5]);
    printf ("  %f, %f, %f]\n", c[6], c[7], c[8]);
    return 0;
  }
```

The code in Listing 15.1 can be compiled and run with:

```
$ gcc example.c -I/path/to/blas/include -L/path/to/blas/lib -lgsl
-lgslcblas
$ ./a.out
[ 0.300000, -0.100000, 0.000000
  0.200000, 0.000000, 0.000000
  0.700000, -0.500000, 0.000000]
```

In the compilation line above, the `/path/to/blas` strings should be replaced with the path to BLAS on your machine.

The three examples of BLAS routine calls shown in this section aim to show how straightforward it is to use existing BLAS routines for performing linear algebra. The key takeaway, whenever performing linear algebra in your software, is to find and use a BLAS library that can be included in your project, saving you the trouble of writing these routines yourself.

15.4 LAPACK

Based on the BLAS routines, the LAPACK routines offer higher mathematical functionality, such as linear system solving and eigenvalue calculations.

For a simple example, `dgesv` is the double-precision (`d`) general matrix (`ge`) linear system solver (`sv`). You give this routine a matrix A and a right-hand side b, and it returns the vector x that satisfies $Ax = b$.

Let's look at this routine, since it illustrates a number of points already covered, as well as some new ones.

```
call dgesv(n, nrhs, a, lda, ipiv, x, ldb, info)
```

You see that the matrix a is specified by the parameters n, lda as explained above. The vector b is of size n, but there is a further parameter nrhs: you can actually solve more than one linear system at a time. If nrhs is more than one, there is a further parameter, ldb, specifying the leading dimension of *b*.

Next you will notice that there is actually no parameter for the output vector; instead, the input b is overwritten. In fact, even the matrix a is overwritten after this routine finishes: it will contain the lower upper (LU) factorization of the matrix. That is typical for LAPACK: the library is very parsimonious with temporary space; if you want to save the value of the matrix, you need to make a copy yourself. This is also the reason that the integer array ipiv, which contains the pivots of the factorization, has to be provided by the user, even though the user will most likely never inspect it.

As an example of higher functionality in LAPACK, there are routines like dgecon that estimate the condition number of the matrix, which can be used in error estimation of matrix-based algorithms.

15.4.1 Other Storage Formats

In addition to the general matrices (square and rectangular matrices with consecutive elements in memory), BLAS and LAPACK can manipulate two additional matrix structures. "Packed" matrix formats are used to encode symmetric and Hermitian matrices, which are determined by half of their elements. "Banded" matrix formats are used to encode matrices for which all elements are zero except for a band. That is, for nonzero integers p and q, the matrix elements $a_{i,j} = 0$ where $j < I - p$ and $j > I + q$.

15.4.2 Parallelism

Most processors on the market these days are of a *multicore* design: each processor chip contains several independent processing elements that can divide up a problem because they operate on the same memory. There is a further type of parallelism that is used in *cluster* computers: distributed memory parallelism. For the latter type of parallelism, there are further libraries; here, we mention the Scalapack [2] and Elemental [3] libraries.

Shared memory parallelism is supported by the BLAS and LAPACK libraries, and in fact with the same API. For instance, the Intel MKL library has multithreaded versions of most routines. Such multithreaded routines would typically be used in a single-threaded application that relies on parallel code sections, for instance in library routines, to achieve parallelism. On the other hand, if the user is parallelizing the whole application, and the BLAS routine is called from a parallel thread, then a traditional single-threaded routine should be used. This latter scenario should also be used if the application uses distributed memory programming and the cores are operating independently.

15.5 The PETSc Library for Parallel Linear Algebra

The routines in BLAS and LAPACK are only concerned with dense matrices, that is, matrices that are stored as a rectangular array. However, in certain physics and engineering applications, matrices are *sparse*, meaning that they have so many zero elements that you want to adopt a specialized format to prevent storing the zeroes and operating on them. There have been some efforts to standardize operations on sparse matrices, but none of these have taken hold. Consequently, each sparse matrix package has its own internal format. In this chapter, you will see a library that targets different applications and is of a different design, with smaller routines that are used as a toolbox.

The Portable Extendable Toolkit for Scientific Computing (PETSc) and LAPACK both consider linear systems to be a common concern, but instead of dense systems, it targets mostly sparse systems, such as arising from finite element (FE) calculations. While LAPACK is exclusively for shared memory, PETSc can be used on a single processor or in distributed memory mode through its use of Message Passing Interface (MPI).

Sparse systems of equations are harder to deal with than dense ones for a number of reasons. One is that software is not standardized: there is no sparse equivalent of BLAS or LAPACK. Another is that the irregularity of the problems makes programming them in parallel much more difficult than in the dense case. For such reasons, powerful libraries such as PETSc (notable competitors are Trilinos and Hypre) form the basis of much computational research in computational fluid dynamics (CFD) and related fields.

In this chapter, you will see some basic uses of PETSc, its linear system solving facilities, and some remarks about how the library interacts with other software packages.

15.5.1 What Is in PETSc?

The PETSc library is useful primarily if you have a linear algebra problem to solve that comes from typical engineering applications. For instance, there is support for building coefficient matrices and Jacobians from finite differences and finite elements. Then, there is a large collection of linear and nonlinear system solvers, in particular for the type of sparse systems from finite differences and such. These are all highly customizable.

Additionally, PETSc has support for debugging and profiling, and it contains various utility tools in support of the linear algebra.

The main mode of parallelism is through MPI, but there is support for GPUs and threads.

15.5.2 The PETSc Workflow

The heart of PETSc is a large collection of linear system solvers, mostly for sparse linear systems. In the spirit of a toolbox, none of these routines are monolithic: you assemble and customize the definition of the solver through a number of small library routines. For instance, the linear system solver routine `KSPSolve` requires at least the following calls:

```
KSPCreate(...); // to create the solver object
KSPSetType(...); // to set the type of solver
KSPSetOperators(...); // to set the coefficient matrix
KSPSolve(...); // actual solve call
```

Depending on the type of solver chosen, there are then many other routines to customize the definition.

Unlike in traditional libraries, such as LAPACK, where the user constructs the matrix and vector arrays and passes them to a library routine, in PETSc, that construction is also performed through library routines. A typical workflow for creating a matrix would look like the following:

```
MatCreate(...); // create the matrix object
MatSetType(...); // set the type of matrix
for ( ... i ... )
   for ( ... j ... )
      MatSetValue( ... i j ... ); // fill in matrix value (i,j)
MatAssembly(...);
```

These two sketches show that PETSc is really a *toolkit* that you use to put together your program, rather than a library that is invoked for one specific activity.

15.5.3 Parallelism

PETSc has a very flexible treatment of parallelism that puts only minimal demands on the programmer. For instance, you need to set the type of vectors and matrices to be single processor or parallel, but after that, the rest of the program is independent of that decision. You can even run your parallel program on a single processor and run it with essentially the same efficiency as the uniprocessor program.

Vectors and matrices are parallelized by a partitioning of the column dimension: each process gets a consecutive range of vector indices or matrix rows. For example, a vector of size 100 is divided over 10 processes by giving each a consecutive range of 10 indices. If the number of processes does not evenly divide in the vector size, the excess is spread out in a sensible manner.

It is also possible to indicate the distribution explicitly: the vector size specification looks like

```
VecSetSize( vec, local_size, global_size );
```

where either the local or the global size can be left up to PETSc.

Filling in the elements of a matrix or a vector in parallel takes little care: you want each element to be constructed only once, so a process needs to know which vector elements and matrix rows it owns. For this purpose, calls such as

```
MatGetOwnershipRange(A,&first,&last);
```

exist that return this range.

That makes the matrix creation code:

```
MatGetOwnershipRange(A,&first,&last);
for ( i=first; i<last; i++)
  for ( ... j ... )
    MatSetValue( ... i j ... ); // fill in matrix value (i,j)
```

On the other hand, the fact that a processor *owns* a part of matrix does not imply that it needs to *set* those elements. PETSc is flexible in its handling of parallelism; for instance, if a matrix is read from file, one processor can do the reading and all the setting of matrix elements. The `MatAssembly` calls mentioned above will then move all data in place.

15.5.4 Object Orientation

The design of the PETSc library is inspired by principles from object-oriented programming. That doesn't mean that the library is either written in an object-oriented language or that you would use it from such an environment: PETSc is written in C, and typically, you would use it from C or Fortran. (There is a Python interface, which is briefly discussed in Section 15.5.6.1.) On the other hand, the design clearly uses polymorphism and inheritance.

For instance, when you create a matrix you determine its type, for instance dense or sparse. After that, most methods for the "matrix class" are available in either case, and with identical syntax: the library resolves under the covers what specific routines to call. Cases such as calling an essentially sequential method on a distributed matrix will lead to a runtime error.

15.5.5 Worked-Out Examples

We go through two examples. In the first one, the matrix construction is straightforward. The second example is identical to the first, except that we show how a matrix can be constructed in a finite element context.

15.5.5.1 A Linear System Example

As a first example, let's make a tridiagonal matrix and use it to solve a system of linear equations. Specifically, we will solve the one-dimensional Poisson equation:

$$\begin{pmatrix} 2 & -1 & & & \emptyset \\ -1 & 2 & -1 & & \\ & \ddots & \ddots & \ddots & \\ \emptyset & & & -1 & 2 \end{pmatrix} \begin{pmatrix} x_1 \\ x_2 \\ \vdots \\ x_n \end{pmatrix} = \begin{pmatrix} b_1 \\ b_2 \\ \vdots \\ b_n \end{pmatrix}$$

The matrix has diagonal elements 2 and off-diagonal –1. We construct it as follows:

```
// loop over the matrix elements on this processor
int my_first,my_last;
MatGetOwnershipRange(A,&my_first,&my_last);
for (int row=my_first; row<my_last; row++) {
       // set the diagonal element
       int col = row; double v=2.;
       MatSetValue(A,row,col,v,INSERT_VALUES);
       // set the lower off-diagonal element
       col = row-1; v=-1.;
       if (col>=0) {
         MatSetValue(A,row,col,v,INSERT_VALUES);
       }
       // set the upper off-diagonal element
       col = row+1; v=-1.;
       if (col<number_of_rows) {
         MatSetValue(A,row,col,v,INSERT_VALUES);
       }
}
MatAssemblyBegin(A,MAT_FINAL_ASSEMBLY);
MatAssemblyEnd(A,MAT_FINAL_ASSEMBLY);
```

The loop bounds here are chosen in such a way that each processor only sets the elements that are stored locally. Below, is an example where that is not the case.

We create a right-hand side that is identically one. This is of course only for testing purposes; in practice, you will have a construction loop similar to the one for the matrix.

```
Vec sol,rhs; double one=1.;
VecCreate(comm,&rhs);
VecSetType(rhs,VECMPI);
VecSetSizes(rhs,PETSC_DECIDE,number_of_rows);
VecDuplicate(rhs,&sol);
VecSet(rhs,one);
```

And then, finally, we can solve the linear system:

```
KSP solver;
KSPCreate(comm,&solver);
KSPSetOperators(solver,A,A);
KSPSetFromOptions(solver);
KSPSolve(solver,rhs,sol);
```

Here, we use all default options regarding the choice of what solver we use precisely. Note that PETSc is not intelligent: it will use a default choice, which may work for your particular linear system, or it may not.

Calling a solve routine is not the same as actually solving the linear system: the solver may have failed to reach the required accuracy within the default number of iterations. Since the solver routine returns no further information, we need some more library calls:

```
KSPConvergedReason reason;
KSPGetConvergedReason(solver,&reason);
if (reason<0) {
  PetscPrintf(PETSC_COMM_WORLD,"Failed to converge\n");
} else {
  int iteration;
  KSPGetIterationNumber(solver,&iteration);
  PetscPrintf(PETSC_COMM_WORLD,"Convergence in %d
iterations\n",iteration);
}
```

This fragment illustrates again the toolbox nature of PETSc: since it is impossible to foresee every need of the user, PETSc opts to provide many small routines to construct the user program, rather than a few all-powerful routines.

15.5.5.2 A Finite Element Example

The scheme we used above to construct a matrix will work for many applications, including those from finite difference methods. In the case of finite elements, the story is more complicated. There, a matrix element is the sum of contributions of several finite elements, and these can belong to different processes. We then parallelize over the finite elements, rather than the matrix element. A finite element matrix construction code could look like the following:

```
for ( e=my_first_element; e<my_last_element; e++)
  for ( ... ni ... ) // loop over element nodes
    for ( ... nj ... ) { // loop over element nodes
      I = translate_to_global(e,ni)
      J = translate_to_global(e,nj)
      MatSetValue( ... I,J, ... ADD_VALUES );
```

where the parameter `ADD_VALUES` indicates that the call does not set the matrix element, but contributes to it additively.

15.5.6 PETSc Usage

15.5.6.1 PETSc and Python

The `petsc4py` library (available from https://bitbucket.org/petsc/petsc4py) offers a Python interface to PETSc. As the following incomplete code shows, the interface is more or less a line-by-line translation of the C/Fortran interface. The parameter lists are typically shortened, using judicious defaults.

```
A = PETSc.Mat()
A.create( PETSc.COMM_WORLD )
A.setSizes( [N,N] )
A.setType("mpiaij")

% set elements...

solver = PETSc.KSP()
solver.create( PETSc.COMM_WORLD )
solver.setOperators( A )
solver.solve( b,x )
```

The main advantage of using the Python interface is that this allows PETSc to be seamlessly integrated in an application that, for other reasons, already uses Python. One could, for instance, image using `petsc4py` in a machine learning application, where the matrices were constructed by text analysis on a large e-mail database.

15.5.7 Runtime Options

Choosing an iterative linear system solver often relies more on "numerical folklore" and experimentation than strict mathematical theory. Therefore, it is likely you will need to do some experimenting before you make your definitive choice. PETSc makes such experimentation easy. If before `KSPSetUp` you call `KSPSetFromOptions`, then any commandline options pertaining to `KSP` will override setting made in the code.

For instance, the choice `KSPSetType(solver,KSPCG)`, which selects the Conjugate Gradient method, can be overridden with a commandline option `-ksp_type bicg` to choose the Biconjugate Gradient method.

There are many options, pertaining to matrix type, solver type, and others. The way to find these is to consult the reference information for a routine, which lists the relevant commandline options.

15.5.8 Profiling

PETSc has various profiling tools. The simplest way of profiling your code is to add a commandline option `-log_summary`, which prints out a breakdown of PETSc routines according to their performance by various measures.

For more sophisticated profiling, you can add your own events.

```
PetscLogEventRegister( ... ); // create an event
PetscLogEventBegin( ... ); // indicate event start & end
PetscLogEventEnd( ... );
```

15.5.9 Debugging

PETSc offers debugging assistance in a number of ways. For instance, in case of runtime exceptions, which can be caught, a backtrace of the call tree is displayed. Also, a variety of memory errors can be caught; if you use PETSc's allocation routines, this even extends to user code.

References

1. E. Anderson, Z. Bai, C. Bischof, S. Blackford, J. Demmel, J. Dongarra, J. Du Croz, A. Greenbaum, S. Hammarling, A. McKenney, and D. Sorensen. *LAPACK Users' Guide.* 3rd Ed. Society for Industrial and Applied Mathematics, Philadelphia, PA, 1999.
2. Y. Choi, J.J. Dongarra, R. Pozo, and D.W. Walker. Scalapack: A scalable linear algebra library for distributed memory concurrent computers. In *Proceedings of the Fourth Symposium on the Frontiers of Massively Parallel Computation (Frontiers '92)*, McLean, VA, October 19–21, 1992, pp. 120–127.
3. J. Poulson, B. Marker, J.R. Hammond, and R. van de Geijn. Elemental: A new framework for distributed memory dense matrix computations. *ACM Trans. Math. Softw.*, 39(2), Article No. 13, 2013. http://dl.acm.org/citation.cfm?id=2427030.
4. C.L. Lawson, R.J. Hanson, D.R. Kincaid, and F.T. Krogh. Basic linear algebra subprograms for Fortran usage. *ACM Trans. Math. Softw.*, 5(3):308–323, 1979.

16

Parallel Computing with Accelerators

İnanç Şenocak and Haoqiang Jin

CONTENTS

16.1 Introduction

The last decade has seen several hardware and software technologies proposed to establish a manycore computing paradigm in scientific and technical computing. Some of these technologies found traction and were adopted by thousands of users (e.g., Compute-Unified Device Architecture [CUDA]), and some have fallen out of favor despite huge investments (e.g., the IBM Cell Broadband Engine in Roadrunner, the world's first petascale supercomputer at the Los Alamos National Laboratory). Today, all supercomputers adopt massively parallel, multithreaded architectures. In this chapter, we will show you how to use multithreaded architectures in the context of scientific and technical computing.

The shift to multicore computing started around 2005, when further increases in processor clock frequency resulted in unmanageable levels of heat generation. Since then, clock speeds have remained relatively flat, but the number of cores on chips has increased. Twenty years ago, we would speak of a room-sized 64-node Beowulf cluster with one central processing unit (CPU) per node. Today, an Intel Xeon E7-8890 v3 processor has 18 CPU cores, and modern motherboards can house four of these processors in a server node with a total CPU core count of 72. Intel Xeon Phi accelerators place 61 cores on a single chip, and graphics processing units (GPUs) offer thousands of cores (e.g., NVIDIA's Tesla K40 has 2880 cores). GPUs have a higher core count because these cores are primarily designed for number-crunching and have less control logics, whereas a CPU core is designed to accomplish many tasks, from number crunching to running an operating system.

Advances in manycore hardware need to be matched with new programming models and software technologies to arrive at superior computational performance. To this end, CUDA [1] and Open Accelerators (OpenACC) [2] were developed to better exploit parallelism in GPUs. Open Multi-Processing (OpenMP) 4.0 specification [3], officially released in 2013, provides a model to support both the manycore architecture of Intel Xeon Phi and GPUs. As of August 2015, compiler implementations of the OpenMP 4.0 specification for GPUs were currently underway.

In the following sections, our aim is to introduce the key features of NVIDIA's Compute-Unified Device Architecture (CUDA) and Intel's Many Integrated Core (MIC) architecture. The term *accelerator* is used in a broader sense to refer to both the GPU and the Intel Xeon Phi coprocessors. We will apply the finite difference method to a tractable prototype problem to illustrate the key features of OpenACC, CUDA, and OpenMP and compare their performances (see the CUDA User Guide [1] for more details).

16.2 NVIDIA CUDA Architecture

Modern GPUs are massively parallel computing platforms ideal for fine-grain parallelism, where many identical threads are processed simultaneously. GPUs impress with their high number of cores and high memory bandwidth. NVIDIA's Tesla K40 has 2880 cores with a memory bandwidth of 288 GB/s. A key distinguishing feature of CUDA is its ability to create many threads quickly with little overhead. The number of threads on a GPU can greatly exceed the number of cores. GPU threads are lightweight compared to CPU threads because a GPU thread does not have to store its state like a CPU thread does. CUDA gives the programmer precise control over the number of threads and the organization of memory on the GPU; it allows a skilled programmer to utilize the GPU hardware optimally for a particular program. Doing so requires detailed knowledge of how the threads are scheduled for execution and the latencies associated with each type of memory (see CUDA User Guide for more details), but certain rules-of-thumb can also help achieve good performance.

CUDA introduces a *thread hierarchy*, allowing a multithreaded program to be expressed as a grid of *thread blocks*. Each block of thread is then scheduled behind the scenes to be processed on a streaming multiprocessor (SM) as space become available. This feature is key to the scalability of computations on a GPU. An SM is a group of cores on a GPU put together to process a *thread block*. The 2880 cores of Tesla K40 are distributed over 15 SMs, each having 192 cores. In addition to thread blocks, a collection of 32 threads is called a *warp* in CUDA. A warp calculates a single instruction in lockstep on an SM. Another concept is the *half-warp* in which global memory accesses can be coalesced in a single transaction. Therefore, it is beneficial, performance wise, to create thread blocks that are multiples of 16 threads. Note that the maximum number of threads per multiprocessor is 2048 for Tesla K40, and the best practice for performance is to have 256 or 512 threads per thread block.

In addition to a thread hierarchy, CUDA provides a *memory hierarchy*. There are separate memory spaces that can be fast or slow with different read and write permissions. The *global memory* on the device can be quite large (e.g., 12 GB on Tesla K40). It can be read and written from both the host (i.e., CPU) and the device (i.e., GPU). All threads have access to global memory, but access can be slow. Therefore, coalescing the accesses help hide the latency.

Shared memory is fast, but small in size. Only the threads within the same block have direct read and write capabilities to the shared memory. In NVIDIA's Kepler architecture, programmers can adjust the size of the shared memory. By default, there is 64 KB of on-chip memory that can split between the *L1 cache* and the shared memory on devices with *compute capability* 3.0 and higher. By default, there is 48 KB of shared memory per block.

Register memory is private to a thread. There are 64,000 registers per multiprocessor and 255 per thread on devices of compute capability 3.0 and higher. Registers are very fast memory. Variables declared locally in the kernel are stored on registers. When there are not enough registers, which is referred to as "register pressure," registers can spill over to the *local memory*, which is also private to a thread but stored in the device memory. Therefore, access will be slow if local memory is not cached in L1. Register pressure can also be alleviated by using the *constant memory* for read-only variables in the kernel. Constant memory is cached and can be read and written by the host. Constant memory, as the name implies, is ideal for constant parameters in a computation. The total size of constant memory is 64 KB.

The last type of memory that is available in CUDA is *texture memory*, which can be advantageous for some applications. Texture memory is a read-only memory that is accessible by all threads. It is cached on a chip. It can be advantageous for applications where memory access pattern has spatial locality.

CUDA provides many features to implement various parallel algorithms efficiently. For an in-depth and thorough introduction to the subject, we refer the reader to excellent books written on CUDA programming by Ruetsch and Fatica [4], Sanders and Kandrot [5], and Kirk and Hwu [6], which we have benefited from enormously over the years.

16.3 Intel Many Integrated Core Architecture

The Many Integrated Core (MIC) architecture from Intel is another example of modern accelerator technology that pushes the envelope for parallel computing. It combines many smaller, lower-power Intel processor cores onto a single chip to achieve a high degree of parallelism and better power efficiency. One advantage of the MIC-based accelerators over GPU accelerators is that the Xeon Phi has extended the ×86 instruction set to support wide vector processing units and uses standard parallel programming models, such as message passing interface (MPI), OpenMP, and hybrid MPI + OpenMP, which makes code development simpler. The current generation of the Xeon Phi accelerator (Knights Corner [KNC]) is attached to the host processor via PCI Express bus on a heterogeneous node. The host processor drives both operation and computation of the accelerator. The next generation of Xeon Phi (Knights Landing [KNL]) offers a path to operate as a standalone processor and will be binary-compatible with ordinary Xeon processors.

A typical KNC accelerator contains 60+ cores running at a lower clock speed (1.05–1.24 GHz) for less power consumption. Each KNC core is capable of four hardware threads, but a thread on a KNC core can only issue instructions every other cycle. Thus, a minimum of two threads per core is needed to fully utilize the resource. Each core supports 512-bit-wide vector instructions, compared to 256-bit unit for a host processor, and has a 512 KB unified L2 cache. Each core can deliver 16 double-precision floating-point operations per cycle. Memory on a chip is limited to 16 GB and is not shared with the host processors. Communication with the host processor has to go through the PCI Express bus.

The Xeon Phi accelerator runs on a lightweight, micro Linux operating system (Intel Manycore Platform Software Stack [MPSS]). Intel provides compilers and libraries, including the support of OpenMP and MPI, for application development [7]. There are essentially two modes for programming and execution of applications for Xeon Phi accelerators:

- *Offloading mode*: An application with highly thread-parallel code segments is annotated with additional "offload" directives, such as device constructs in OpenMP 4.0. The code is compiled as an ordinary OpenMP program. The executable is launched on the host and the compute-intensive segments are offloaded to the Xeon Phi accelerators, including data transfer between the host and the accelerators.
- *Native mode*: Programming for the accelerator is done in standard C/C++ and Fortran languages. The same program source code written for the standard Intel Xeon processor can be compiled for the Xeon Phi accelerator to run natively. This mode is often used in conjunction with MPI to run codes on both host processors and accelerators in a symmetric mode for full resource utilization.

16.4 Two-Dimensional Wave Equation

We will solve the two-dimensional (2D) wave equation with a finite difference method to illustrate parallel computing with OpenACC, OpenMP, and CUDA. Let's consider a rectangular membrane (e.g., a drumhead) 4 ft × 2 ft wide in x and y directions, respectively. The membrane is homogeneous and perfectly flexible and offers no resistance to bending. The displacement of the membrane from rest at a point (x, y) is governed by the following second-order linear partial differential equation (PDE):

$$\frac{\partial^2 \phi}{\partial t^2} = c^2 \left(\frac{\partial^2 \phi}{\partial x^2} + \frac{\partial^2 \phi}{\partial y^2} \right), \tag{16.1}$$

where ϕ is the wave height, c^2 is equal to 5.0, x and y are the spatial coordinates, and t is the time. Note that c^2 represents membrane tension T divided by the membrane mass per unit area ρ. The two initial conditions are as follows:

$$\phi(x, y, t) = 0.1(4x - x^2)(2y - y^2), \tag{16.2}$$

$$\left. \frac{\partial \phi}{\partial t} \right|_{t=0} = 0. \tag{16.3}$$

The membrane is clamped at all the edges, meaning ϕ is zero along all the edges.

We use a uniform mesh such that $\Delta x = \Delta y = h$ and second-order accurate central difference schemes to discretize the temporal and spatial derivatives in Equation 16.1 to arrive at the following discretized form of the wave equation:

$$\phi_{i,j}^{t+1} = 2\phi_{i,j}^{t} - \phi_{i,j}^{t-1} + \frac{c^2 \Delta t^2}{h^2} \left(\phi_{i+1,j}^{t} + \phi_{i-1,j}^{t} + \phi_{i,j+1}^{t} + \phi_{i,j-1}^{t} - 4\phi_{i,j}^{t} \right), \tag{16.4}$$

where Δt is the time step size [8].

Note that the right-hand side depends on the values from the current (t) and old ($t-1$) time steps. From the discretization of the initial condition given in Equation 16.3 with a central difference formula in time, we have the following:

$$\frac{\phi_{i,j}^{t+1} - \phi_{i,j}^{t-1}}{2\Delta t}. \tag{16.5}$$

Substituting $\phi_{i,j}^{t+1}$ from Equation 16.5 into Equation 16.4, we obtain the value at $t-1$ as follows:

$$\phi_{i,j}^{t-1} = \phi_{i,j}^{t} + \frac{c^2 \Delta t^2}{2h^2}\left(\phi_{i+1,j}^{t} + \phi_{i-1,j}^{t} + \phi_{i,j+1}^{t} + \phi_{i,j-1}^{t} - 4\phi_{i,j}^{t}\right). \tag{16.6}$$

With the initial conditions at time t (Equation 16.2) and $t-1$ (Equation 16.6), we can use Equation 16.4 to advance the solution in time. Note that Equations 16.2 and 16.6 are only used in the first time step to initiate time advancement. Equation 16.4 is solved on a computational mesh of 8192 × 4096 for all the cases in this chapter. Because we use an explicit scheme we ensure that the time step and grid size do not violate the Courant–Friedrichs–Lewy (CFL) condition:

$$\frac{c\Delta t}{h} < 1.0. \tag{16.7}$$

Analytical or exact solutions are very helpful in ensuring correctness in a numerical implementation. The following analytical solution to the 2D wave equation is available [9]:

$$\phi(x,y,t) = 0.426050 \sum_{m=1,\text{odd}}^{\infty} \sum_{n=1,\text{odd}}^{\infty} \left(\frac{1}{m^3 n^3} \cos\left(t\frac{\sqrt{5}\pi}{4}\sqrt{m^2+4n^2} \right) \sin\left(\frac{m\pi x}{4} \right) \sin\left(\frac{n\pi y}{2} \right) \right). \tag{16.8}$$

16.5 Serial C Implementation for CPUs

We use C programming language to implement the numerical solution of the 2D wave equation using the finite difference method. We will not explain the computer implementation of the entire program here because of space limitations; however, the computer program is straightforward and available through a repository with plenty of comments. We will restrict our discussion to sections of the serial code that are most relevant to the parallel implementation.

Listing 16.1 shows the dynamic memory allocation of the arrays `unew`, `u`, and `uold` on the CPU, which is common to all parallel implementations. Pointers to these memory spaces are declared as we are going to make use of pointer arithmetic. We created the function `solveWave` as shown in Listing 16.2b, which implements the finite difference formula given in Equation 16.4. Because we prefer to flatten a 2D array into a 1D array in our implementation for ease of coding, we define macros for indexing (e.g., IC for [i][j], IM1

```
175     REAL *unew,  *u,     *uold, *tmp;
176
177     unew              = (REAL *)calloc(NX*NY,sizeof(REAL));
178     u                 = (REAL *)calloc(NX*NY,sizeof(REAL));
179     uold              = (REAL *)calloc(NX*NY,sizeof(REAL));
```

LISTING 16.1
Dynamic memory allocation for the serial, OpenMP, and OpenACC versions of the 2D wave equation example.

```
189 #pragma acc data pcopyin    ( unew[0:NX*NY], uold[0:NX*NY] )  \
190                  pcopy      ( u[0:NX*NY]     )
191 {
192     for (INT n=1; n<nTimeSteps+1; n++)
193     {
194         solveWave( unew, u, uold );
195
196         tmp  = uold;
197         uold = u;
198         u    = unew;
199         unew = tmp;
200     }
201 }
```
(a)

```
58 void solveWave (REAL *RESTRICT unew, const REAL *RESTRICT u, const REAL *RESTRICT uold)
59 {
60     INT i,j;
61 #pragma acc kernels loop /*comment this line to use the parallel construct */
62 //#pragma acc parallel loop gang vector_length(256) /*uncomment for parallel construct*/
63     for (j= 1; j<NY-1 ; j++)
64     {
65         #pragma acc loop vector
66         for (i=1 ; i<NX-1 ; i++)
67         {
68             unew[IC] = 2.0f*u[IC] - uold[IC] +
69                        FACTOR*( u[IP1] + u[IM1] + u[JP1] + u[JM1] - 4.0f*u[IC] );
70         }
71     }
72 }
```
(b)

LISTING 16.2
(a) The main program where data are offloaded to the GPU with OpenACC directives. (b) OpenACC parallelization of function `solveWave` for the finite difference formulation in Equation 16.4.

for [i–1][j], and JP1 for [i][j+1]) in the source code. Function `solveWave` is then advanced in time by calling it in the `main` function (line 194) within a loop, as shown in Listing 16.2a.

Pointers to the memory spaces are being swapped after each time iteration (lines 196–199), instead of copying the individual elements of an array, which would be inefficient because of data movement. Note that pointer swapping in the time advancement loop is common to all versions of the parallel codes presented in this chapter. It is especially beneficial to achieve good performance on accelerators because data are kept local to the device.

16.6 Parallel Implementation with OpenACC for GPUs

OpenACC emerged in 2011 as a high-level programming model for accelerators (i.e., GPUs). It provides parallelism through compiler directives similar to OpenMP. The promise of

OpenACC is its portability; it is not restricted to a specific accelerator. Currently, it supports GPUs from NVIDIA and AMD corporations. To maintain portability into the future, OpenACC supports multiple levels of parallelism and a memory hierarchy. OpenACC provides directives to offload data and computation from a host device to an accelerator. In a typical situation, the host would be the CPU and the accelerator would be the GPU. CPU and GPU may have different memory spaces; therefore, OpenACC provides directives to manage data movement between the host and the accelerator. However, unlike the common practice in CUDA to declare separate pointers to manage different memory spaces, OpenACC can manage different memory spaces through a single declaration of pointers, which eases parallel programming, debugging, and code maintenance tasks significantly. Programmers who are experienced with OpenMP will find OpenACC programming familiar. In our wave equation example, the serial code remained unchanged, and a few lines of directives had to be added for parallel computing on the GPU. Therefore, it would be fair to say that OpenACC is less intrusive in terms of porting a serial code to accelerators than CUDA is.

The common syntax of OpenACC is the keyword `acc` followed by a directive and optional clauses. In C/C++, the directives are represented by `#pragma`. The keyword `acc` informs the compiler that the directives that follow are from the OpenACC application programming interface (API). The directives can be applied to a block of code contained by curly braces ({...}).

Porting a serial scientific simulation code to a parallel computing platform can be a daunting task. A systematic effort is needed to identify compute-intensive sections of a code. Because of scope and page limitations, we will not delve into the details of such labor-intensive efforts. As a best practice for parallel computing on accelerators, a programmer should aim to minimize peripheral component interconnect express (PCIe) data transfers between the host and the accelerator. It is difficult for a compiler to detect data locality. Therefore, it is the programmer's responsibility to keep data local and copy data back to the host only when needed. In the finite difference solution of the wave equation, we need the wave amplitude φ at time $t + 1$, t, and, $t - 1$ (i.e., variables `unew`, `u`, and `uold`). In the function `solveWave`, `unew` is updated at each time step on the accelerator, and pointers are then swapped before moving on to the next time step. Therefore, it is beneficial, performance wise, to offload these data to the accelerator once and only copy them back when needed. OpenACC provides a `data` region clause to achieve this. The use of the `data` region is shown in Listing 16.2a at line 189. Note that the `data` region is outside of the time advancement loop to keep data local on the accelerator during time marching of the solution.

Data clauses such as `copy`, `copyin`, and `copyout` gives more control to the programmer to assert data locality. The `copy` clause allocates space for a variable and copies it onto the device and copies the results back to the host at the end of the `data` region. The `copyin` does the same operations as `copy` except copying the results back to the host. The `copyout`, on the other hand, allocates space but only copies the data back to the host. Because data movement between the host and the device are expensive, a programmer should benefit from these three operations by better understanding the use of data in their algorithm.

OpenACC also enables further control over the data. The `present` clause informs the compiler that data are already present on the device. The `create` clause, as the name suggests, creates space for the variables but does not perform `copyin` or `copyout`. In our example, we use the `pcopy` and `pcopyin` to comply with future specifications of OpenACC. `pcopy` and `pcopyin` are abbreviated versions of the `present_or_copy` and `present_or_copyin` clauses, respectively, which combine the functionality of different data clauses into a single clause. For instance, with the `pcopy` clause, the compiler checks if the variable is present on the device, if it is available, it uses the existing copy, and if not, it will copy the variable to the device. With C/C++ compilers, OpenACC also needs to know the size of the arrays; therefore, we provide that information within the data region. For `unew`, `u`, and `uold`, we specify the start index and the number of elements as `[0:NX*NY]` on lines 189 and 190 in Listing 16.2a. Note that, depending on the algorithm, we can also transfer a portion of the data by specifying the start index and the number of elements.

It is easy to identify the highly parallelizable compute-intensive section in our example code, which is the `solveWave` function, because of the explicit finite difference formulation. For complex codes, a profiler would be beneficial to achieve this task. OpenACC provides the `parallel` and `kernels` constructs to parallelize loops. We will consider both approaches to parallelize the nested loop in function `solveWave`.

The `kernels` construct relies on the automatic parallelization capabilities of the compiler. This construct is ideal for loops and nested loops where there is explicit parallelism. With this construct, the compiler is free to best map the loop to the parallel architecture of the accelerator. It is also up to the compiler to parallelize the code. The programmer is only giving a hint by assigning loops to a kernels construct. The compiler may choose not to parallelize the loop. Listing 16.2b shows the use of the `kernels` construct at line 61. We also used the loop clause on line 65 as an optimization to tell the compiler that the inner loop is a parallelizable vector loop.

With the `parallel` construct, the `loop` clause is necessary to inform the compiler that it is safe to parallelize the loop. It is the programmers' responsibility to assert parallelism in loops to the compiler. If the programmer misidentifies parallelism and data dependency, the compiler will proceed and parallelize the loop, leading to incorrect results. The `parallel` directive needs to be applied to each loop separately. A block of code with multiple loops cannot be assigned to a single `parallel` construct. In Listing 16.2b, line 62 shows the use of the `parallel` construct. Line 65 is again retained to tell the compiler about the inner vector loop. The optional clauses on the parallel construct, such as the `vector_length()`, inform the compiler to make better use of the hardware. Based on a CUDA best practice of using 256 threads per CUDA block, we found that using a vector length of 256 in OpenACC on NVIDIA GPUs provides roughly 5% improvement in speed over the default OpenACC directive (128 threads) where no vector length is specified explicitly. The optional `gang` clause informs the compiler that the outer loop is a gang loop. We let the compiler decide the number of gangs in this case. The optional `num_gangs()` clause enables the programmer to specify an integer number for gangs.

OpenACC provides three levels of parallelism through *gang*, *worker*, and *vector*. Fine-grained Single Instruction Multiple Data (SIMD) parallelism is supported by vector parallelism, whereas gang parallelism addresses the coarse-grained parallelism. Each gang is independent and composed of one or more workers, where each worker operates on a vector of some size. Because each gang is independent, they may not synchronize among themselves. However, synchronization within a single gang is possible, and vectors and workers can share a common cache on a gang. To this end, a gang resembles to a CUDA block. Additionally, OpenACC provides the `seq` clause to force sequential processing where parallelism is not desired.

Note that on line 58 in Listing 16.2b, we use the `restrict` keyword, which is very helpful for the OpenACC compiler to detect parallelism. Using the `const` keyword also tells the compiler to make the best use of the memory hierarchy on the accelerator. For instance, the compiler may choose to use the read-only memory on the accelerator.

In this section, we have restricted our discussion to a few OpenACC directives and clauses to parallelize the 2D wave equation program. Further information is available in the OpenACC specification and best practices guide, including advanced features to enable parallelization of complex scientific simulation codes.

16.7 Parallel Implementation with OpenMP for Intel Xeon Phi

OpenMP is a high-level, directive-based model for parallel programming on shared-memory systems. Support for accelerators, such as Intel Xeon Phi, has been added in the 4.0 release of OpenMP. Similar to OpenACC, the model provides language directives to annotate segments of code and data to be offloaded to an accelerator for execution in a host-centric paradigm—the host drives both computation and data movement. OpenMP provides the `target` construct for specifying an offloading code region and the `map` clause on the `target` directive for data mapping between the host and the device. Mapping involves the creation of a device copy of the data if the data do not exist on the device and the transfer of the data content between the host and the device. The `declare target` directive can be used to map global data to the device.

As an example, the solver routine of the wave code annotated with OpenMP directives is illustrated in Listing 16.3b. The `target` directive on line 57 specifies that the code segment between lines 59 and 69 will be offloaded to the device for execution. The `to` modifier in the `map` clause specifies arrays `u` and `uold` of size `NX*NY` starting from the lower bound 0 to be

```
204 #ifdef OFFLOAD
205 #pragma omp target data map( to: unew[0:NX*NY], uold[0:NX*NY] ) \
206                         map( tofrom: u[0:NX*NY] )
207 #endif
208  {
209     REAL *tmp;
210     INT n;
211     for (n=1; n<nTimeSteps+1; n++)
212     {
213        solveWave( unew, u, uold );
214
215        tmp  = uold;
216        uold = u;
217        u    = unew;
218        unew = tmp;
219     }
220  }
```

(a)

```
52 void solveWave ( REAL *RESTRICT unew, REAL *RESTRICT u, REAL *RESTRICT uold )
53 {
54     INT i,j;
55
56 #ifdef OFFLOAD
57 #pragma omp target map( to: uold[0:NX*NY], u[0:NX*NY] ) \
58                    map( from: unew[0:NX*NY] )
59  {
60 #endif
61 #pragma omp parallel for private(i,j)
62     for (j=1; j<NY-1; j++) {
63        for (i=1; i<NX-1; i++) {
64           unew[IC] = 2.0f*u[IC] - uold[IC] +
65                      FACTOR*( u[IP1] + u[IM1] + u[JP1] + u[JM1] - 4.0f*u[IC] );
66        }
67     }
68 #ifdef OFFLOAD
69  } // end-target
70 #endif
71 }
```

(b)

LISTING 16.3

(a) The main program where data are offloaded to the Intel Xeon Phi accelerator with OpenMP directives.
(b) OpenMP parallelization of function `solveWave` for the finite difference formulation in Equation 16.4.

copied from the host to the device at the beginning, and the `from` modifier specifies array `unew` of size `NX*NY` to be copied back from the device at the end after computation. By default, the mapped data live only within the scope of the `target` region; that is, after code execution, the mapped data (`u`, `uold`, `unew`) will be deleted from the device. The parallel execution on the device is governed by the `parallel for` directive on line 61, as defined by the usual OpenMP fork-join model. Without the `target` directive, the code is just a regular OpenMP code and can be run on either the host processor or the Xeon Phi accelerator natively. The `restrict` keyword for the function arguments informs the compiler that there is no pointer aliasing between the arguments. This allows the compiler to exploit vector parallelism for the inner `i` loop at line 63, which is very important for performance, especially on Xeon Phi.

The major overhead in the offload model is the cost of data transfer between the host and the device. To remedy the situation, OpenMP provides a few options to reduce the transfer cost. One option is to use the `target data` construct to define a coarser data region on the device in which multiple `target` constructs can use the previously mapped data without data remapping and transfer. If necessary, one can use the `always` modifier to the `map` clause or the `target update` construct to force data transfer. The second option for reducing the overhead is to overlap data transfer (via the `target update` construct) with host computation by wrapping them inside asynchronous tasks. The `taskwait` directive or task dependences can then be used to perform task synchronization. For readers who are interested in this topic, please refer to other references, such as advanced OpenMP tutorial given by Terboven et al. [10].

The code snippet in Listing 16.3a illustrates the use of the `target data` construct for the time step loop that calls the solver routine multiple times. Data are copied once for arrays (`u`, `uold`, `unew`) from the host to the device before the time loop. Within the time loop, the solver routine in `solveWave()` is offloaded by the `target` construct to the device for execution without data remapping of the mapped arrays. At the end of the `target data` region, array `u` is copied back from the device to the host.

16.8 Parallel Implementation with CUDA for GPUs

It would be fair to say that CUDA, which debuted in 2006, has been a game-changing technology and a catalyst for several efforts in parallel computing. GPUs have long been appreciated for their potential for parallel computing, but general purpose computing on GPUs has been a difficult task for scientists and engineers who were not experts in computer graphics programming. Prior to the introduction of CUDA, scientific computing on GPUs was a tricky exercise using either OpenGL or DirectX due to limitations in the ability to process floating-point data and write at arbitrary memory locations on the GPU. CUDA enabled general purpose computing on GPUs by supporting read and write operations to arbitrary memory locations and complying with the IEEE 754 standard for floating-point calculations for single- and double-precision data (i.e., for compute capability 2.0 and higher).

Unlike OpenMP and OpenACC directives, CUDA is a language extension. Today, we have CUDA extension to the C and Fortran programming languages. In this chapter, our focus is on CUDA C. We will cover a few essential CUDA features to get started with scientific computing on GPUs. For further usage, such as "Professional CUDA C Programming" by Cheng and Grossman dedicated to CUDA programming for a complete introduction to the API.

Because a GPU has its own memory space, CUDA provides functions to copy data back and forth between the host (CPU) and the device (GPU). In our wave equation example, we need to copy the `unew`, `u`, and `uold` to the device. To do this, we first declare the device pointers as shown on line 209 in Listing 16.4a. For programming clarity, we attach the prefix `d_` to pointers that point to device memory. We call `cudaMalloc()` to allocate space for `d_unew`, `d_u`, and `d_uold` on the device, as shown on lines 210–212 in Listing 16.4a. After this step, we copy the content of the host side arrays `unew`, `u`, and `uold` to the global memory on the device using `cudaMemcpy` with the parameter `cudaMemcpyHostToDevice`, as shown on lines 216–218 in Listing 16.4a. Note that, in OpenACC, these data offloading steps and separate pointers to device and host memory are not needed. To avoid repeated PCIe bus data transfer, which is expensive, we have allocated memory on the device and copied their contents once outside of the time marching loop.

```
208 /* create device copies of unew, u, uold */
209     REAL *d_unew, *d_u, *d_uold, *tmp;
210     cudaMalloc( (void**) &d_unew,  NX*NY*sizeof(REAL) );
211     cudaMalloc( (void**) &d_u,     NX*NY*sizeof(REAL) );
212     cudaMalloc( (void**) &d_uold,  NX*NY*sizeof(REAL) );
213
214     initWave( unew, u, uold, x, y );
215
216     cudaMemcpy( d_u, u,           NX*NY*sizeof(REAL), cudaMemcpyHostToDevice );  //memcpy(dest,src,...
217     cudaMemcpy( d_uold, uold, NX*NY*sizeof(REAL), cudaMemcpyHostToDevice );  //memcpy(dest,src,...
218     cudaMemcpy( d_unew, unew, NX*NY*sizeof(REAL), cudaMemcpyHostToDevice );  //memcpy(dest,src,...
219
220 /*  set up the GPU grid/block model */
221
222     dim3 dimGrid  ( N_BLOCKS_X , N_BLOCKS_Y  );
223     dim3 dimBlock ( N_THREADS_X, N_THREADS_Y );
224
225     for (INT n=1; n<nTimeSteps+1; n++) {
226
227       solveWaveGPU <<<dimGrid,dimBlock>>>(d_uold, d_u, d_unew);
228
229       tmp    = d_uold;
230       d_uold = d_u;
231       d_u    = d_unew;
232       d_unew = tmp;
233
234     }
235
236     cudaMemcpy( uFinal, d_u, NX*NY*sizeof(REAL), cudaMemcpyDeviceToHost );  //memcpy(dest,src,...
```

(a)

```
62 __global__ void solveWaveGPU ( REAL *RESTRICT unew, const REAL *RESTRICT u,
63                                const REAL *RESTRICT  uold)
64 {
65
66      INT i,j;
67
68      i = blockIdx.x*blockDim.x + threadIdx.x;
69      j = blockIdx.y*blockDim.y + threadIdx.y;
70
71      if (i>0 && i < (NX-1) && j>0 && j < (NY-1) ) {
72
73         unew[IC] = 2.0f*u[IC] - uold[IC] +
74                    FACTOR*( u[IP1] + u[IM1] + u[JP1] + u[JM1] - 4.0f*u[IC] );
75      }
76
77 }
```

(b)

LISTING 16.4

(a) The main program where device memory is allocated and the CUDA kernel is launched. (b) CUDA kernel `solveWaveGPU` for the finite difference formulation in Equation 16.4.

For the next step, we have to inform CUDA about execution configuration by setting the dimensions of the CUDA grid (line 222) and the CUDA blocks (line 223). We use a block size of 32 × 16, since NVIDIA GPUs with compute capability 3.5 supports up to 1024 threads per block and 2048 per multiprocessor. We are free to choose a different block size, such as 16 × 16. It is recommended that programmers experiment with different block sizes to fine tune the performance. We followed the best practice guidance as mentioned in Section 16.2. In our case, 32 × 16 (512 threads) executed slightly faster than 16 × 16 (256 threads). Grid dimensions are then calculated based on the finite difference mesh size N_BLOCKS_X = NX / N_THREADS_X and N_BLOCKS_Y = NY / N_THREADS_Y. Note that the maximum grid dimensions is 2,147,483,647 × 65,535 × 65,535 in the x, y, and z directions, respectively.

We then call our kernel `solveWaveGPU`, as shown on line 227, with an angle bracket syntax (<<<...>>>). In CUDA terminology, functions that execute on the GPU are called *kernels*. Pointers to the device memory are swapped on the host side after each time iteration, which is very similar to the way we programmed in the serial version. Note that we keep the data on the device by swapping the pointers.

In kernel `solveWaveGPU`, shown in Listing 16.4b, we add the __global__ identifier to the function to tell the compiler that this function will execute on the GPU. Note that the __global__ identifier has nothing to do with the global memory on the device. We kept the finite difference formula (lines 73–74) unchanged from the serial version. The biggest difference is the elimination of `for` loops. Instead, we use the predefined variables `blockIdx`, `blockDim`, and `threadIdx` to find the global array index `i` and `j` as shown on lines 68 and 69 in Listing 16.4b. `blockIdx` gives the index of a block within a grid and `blockDim` contains the number of threads in a block. The index of a thread within a block is obtained through the `threadIdx`. We then use an `if` statement to give a range of index for threads to process simultaneously in a Single Instruction Multiple Data (SIMD) fashion.

At the end of the time marching, we copy the solution stored in `d_u` to an array `uFinal` on the host using `cudaMemcpy` with the parameter `cudaMemcpyDeviceToHost`, as shown on line 236 in Listing 16.4a. Although not displayed in the Listing, we call `cudaFree()` to free up the allocated memory for `d_unew`, `d_u`, and `d_uold` before exiting the main function.

It is not relevant to our simple finite difference example, but let's assume that a different algorithm might need to operate on `unew`. Unlike serial computing, we cannot simply use `unew` in kernel `solveWaveGPU` for a different calculation within the same kernel, because CUDA does not allow synchronization of independent blocks within a kernel. We have no control of knowing when blocks are done updating the global memory. Therefore, we would need to launch a separate kernel after the `solveWaveGPU` kernel to operate on the updated `unew` values. In other words, a kernel launch acts as an implicit global synchronization of all CUDA blocks. It is not uncommon to overlook this global synchronization issue as one transitions from serial computing to parallel programming with CUDA. We should mention that `__syncthreads()`, which can be used inside a kernel, enables threads within the same block to synchronize. It is not meant for global synchronization.

16.9 Performance Discussion

It is difficult to claim a single programming language as the best one. Every language has its own advantages and disadvantages for an application. Sometimes, we simply prefer a particular language because we have a deep knowledge base. Likewise, we don't think we can claim a best programming model for parallel computing. The choice eventually depends on the existing code base, the expertise in the software development team, and available hardware resources. Our goal in this chapter was to introduce the reader to the current state in parallel computing with different accelerators using a tractable prototype problem.

By comparing the code snippets in all listings, we can clearly see that both OpenMP and OpenACC allow the serial code to remain unchanged and require the least amount of coding. All we needed was three lines of OpenACC directives to accelerate the serial code by a factor of 19, which is impressive. The nonintrusive nature of both OpenMP and OpenACC may be advantageous for legacy codes. Although more lines of code needed to be written for the CUDA version, the changes to the serial code were not major. The majority of the serial code remain unchanged.

Figure 16.1 shows strong scaling results on a conventional multicore Intel processor. Using OpenMP on a motherboard with dual 8-core Intel Xeon E-2670 processors (total of 16 cores), we did not gain any speedup beyond 8 cores. The gain beyond 4 cores was also not impressive either. The parallel efficiency of 4 cores was 96%, and it went down to 61% at 8 cores. At 16 cores, the efficiency was 29%. The drop in parallel efficiency on conventional multicore processors is one of the reasons why the computer industry has invested in manycore platforms. Intel Xeon Phi KNC 5110P has 61 physical cores and supports up to 244 threads. As we see from Figure 16.1, there is no speedup beyond 60 threads. OpenMP provides an OFFLOAD mode option to support accelerators. Figure 16.1 shows that the OFFLOAD mode option is nearly twice as slow compared to the native mode. It is likely that the implementation of the `target` construct in the Intel C compiler is creating an extra overhead, which may improve in future versions.

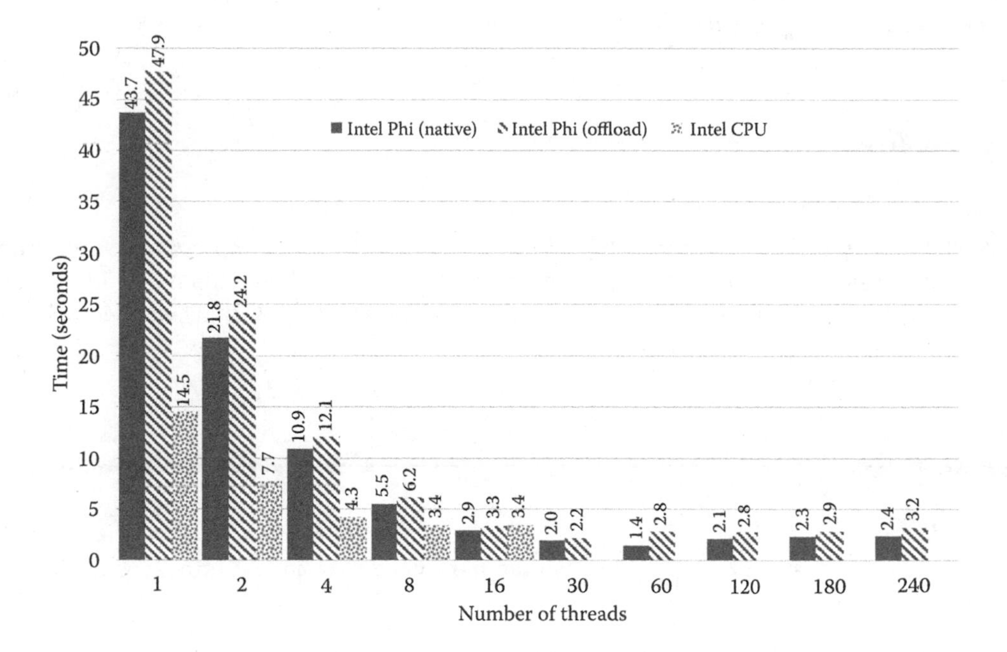

FIGURE 16.1
Timing of the 2D wave equation on Intel Xeon E-2670 (dual 8-core, 2.6 GHz) and Intel Xeon Phi KNC 5110P (61 cores, 1.052 GHz) as a function of threads used in the computations.

TABLE 16.1

Timing of Serial and Various Parallel Implementations of the 2D Wave Equation and Speedup Relative to the Serial Single Core Performance on Intel Xeon E-2670 2.6 GHz 8-Core Processor Using Intel C/C++ Compiler Version 2015.0.090

	CUDA	OpenACC	OpenMP Intel Phi (60 Threads Native)	OpenMP Intel Phi (60 Threads Offload)	OpenMP CPU (8 Threads)	Serial CPU
GTX Titan	0.815 s 17.8×	0.763 s 19.0×	n/a	n/a	n/a	n/a
Tesla K40	0.979 s 14.8×	0.857 s 16.9×	n/a	n/a	n/a	n/a
Intel compiler	n/a	n/a	1.41 s 10.3×	2.83 s 5.12×	3.41 s 4.25×	14.5 s 1.00×

Table 16.1 presents timing for various combinations. We obtained the best performance using OpenACC on an NVIDIA GTX Titan card, which was approximately 19× faster than the serial CPU implementation and 4× faster than the 8-thread OpenMP performance on Intel Xeon E-2670 processor. Overall, OpenACC on a GTX Titan GPU was 1.85× faster than an Intel Xeon Phi 5110p using the native mode of OpenMP. The difference in performance should be less with the more recent Xeon Phi 7120 accelerator. OpenACC was also faster than CUDA by about 15%. We should emphasize that our CUDA implementation was a basic global memory-only implementation. A better use of the memory hierarchy in CUDA could have led to superior performance.

Although not included in our results, single-precision computations were roughly twice as fast as double-precision computations. We did benefit from using the `restrict` keyword significantly, more so in OpenACC and OpenMP for Intel Xeon Phi. We used `-O3` optimization level to compile all versions of the code.

Acknowledgments

The first author would like to thank Massimiliano Fatica of NVIDIA Corporation for many helpful discussions since the release of the first version of CUDA. This material is based upon work supported by the National Science Foundation under Grant Nos. 1440638 and 1229709.

References

1. NVIDIA Corporation. 2015. CUDA C Programming Guide, version 7.0. http://docs.nvidia.com/cuda/pdf/CUDA_C_Programming_Guide.pdf.
2. OpenACC-Standard.org. 2015. OpenACC Programming and Best Practices Guide. http://www.openacc.org/sites/default/files/OpenACC_Programming_Guide.pdf.
3. OpenMP Application Program Interface, version 4.0, 2013. http://www.openmp.org/.

4. Ruetsch, G., and M. Fatica. 2013. *CUDA Fortran for Scientists and Engineers: Best Practices for Efficient CUDA Fortran Programming*. Amsterdam: Morgan Kaufmann Elsevier.
5. Sanders, J., and E. Kandrot. 2011. *CUDA by Example: An Introduction to General-Purpose GPU Programming*. Upper Saddle River, NJ: Addison-Wesley.
6. Kirk, D.B., and W.W. Hwu. 2013. *Programming Massively Parallel Processors: A Hands-on Approach*. 2nd Ed. Amsterdam: Morgan Kaufmann Elsevier.
7. Overview of Intel Xeon Phi Coprocessor. 2015. https://software.intel.com/en-us/mic-developer.
8. Pletcher, R.H., J.C. Tannehill, and D. Anderson. 2013. *Computational Fluid Mechanics and Heat Transfer*. 3rd Ed. Boca Raton, FL: CRC Press.
9. Kreyszig, E. 2011. *Advanced Engineering Mathematics*. 10th Ed. Hoboken, NJ: John Wiley & Sons Inc. (p. 575).
10. Terboven, C., M. Klemm, and B. de Supinski. 2015. Advanced OpenMP Tutorial—Performance and 4.0 Features. *International Supercomputing Conference (ISC'15)*, Frankfurt am Main, Germany.

17

Testing and Verification

Paul Kwiatkowski

CONTENTS

17.1 Programs Grow and Become Software

Occasionally, computer science faculty will bemoan a mind-set among students best summarized as the "homework mentality": the perception that code is written, run once to receive a grade, and then never seen again. A student might look back fondly on a project because it received a high grade, even if a month later they might not be able to decipher the line-by-line details of their own code. On actual projects, when the humble program grows and becomes living software, this nonchalance is dangerous as mistakes don't simply disappear.

This also manifests in research as a sort of "publishing mentality": a perception that success in publishing a paper justifies any sloppy practices in the code it relies on.

Within long-lived research projects, when code completed in a rush is left messy or buggy it also *does not go away* after the deadline. In fact, the researcher's next project often needs to interact with said messy code. For individuals or teams working on projects built by these successive layers of deadline-driven development, eventually, more time is spent deciphering or fixing the mess than adding to new research.

The single greatest tool to avoid this needless burden is the creation and upkeep of automated tests.

Here, we present the best practices and standards related to testing code, developed over the last several decades of software engineering, and which are gaining similar importance in scientific projects as their dependence on programmatic tools grows.

17.2 Why Test?

It seems obvious that tests, first and foremost, contribute to a sense of correctness, that a program functions as it is intended to. Manual use of the program by a human can verify correctness after it has run, and while this is technically a form of feedback that could be considered a test of the program, the use of the term *testing* in a programming context almost universally assumes automated tests. In cases where testing is run entirely by a person, it is explicitly called out as *manual testing*.

For some, the promise of spending more time on active research and less time mired in a headache of indecipherable code written by a teammate or by one's past self is motivation enough. For those who remain unconvinced, the additional merits of automated tests are presented here.

In software development, a *test suite* refers to all tests related to a specific program or all tests related to an entire project, which might itself be comprised of multiple programs. Rest assured that the benefits of a test suite, of any size, are far-reaching.

17.2.1 Guarding against Regressions

In software, the term *regression* is often used as shorthand for regression bug, a situation where a software feature that was once functioning properly was affected by a change, often an upgrade and often elsewhere within a large system, such that the previously correct feature no longer functions properly. It is worth pointing out that fixing a bug, even a regression bug, is itself another change that might introduce yet more regressions. Programmer humor is rich with references to just how often fixing one bug introduces another.

A robust test suite is one that confirms that even slight modifications to an existing unit of code do not break one or more other units (functions, modules, classes, depending on language) that rely on that newly modified code.

17.2.2 Speed and Confidence

A robust automated test suite is one that allows for quick feedback about the state of all major features and required functionality. Depending on the project, the definition of "quick feedback" may refer to a few seconds, or many minutes. It is quick because, almost without exception, it is faster to run a test suite than to have a human orchestrate tens or hundreds of manual runs of the program to ensure that every detail is correct.

This type of quick feedback is essential for a project to maintain the speed of new work. Manual testing is seductively simple at the dawn of any new project, especially for those not used to writing automated tests. As time goes on, feature lists increase, complexity grows, and the time required for manual testing diverges greatly from the time necessary to maintain and run automated tests.

Any team that has added a testing script or checklist to be manually verified after any changes to their project certainly has their heart in the right place. However, the presence of such a checklist is a sure sign that they are past that point of divergence and would benefit greatly from immediately introducing automated tests.

Done well, tests accrue into a tangible sense of confidence in the project's code from many angles: confidence in results, confidence in making changes, confidence in adding new features, and confidence in any estimated project timeline.

17.2.3 Refactoring

In any long-lived software project, sections of code that represent a given piece of work might move around, be generalized, or as is often said *cleaned up*, which tends to be a catch-all term used to mean *made easier to understand*. In software, the process of this sort of movement, change, and cleanup in the pursuit of easier-to-understand code is called *refactoring*.

For instance, in a simple program that implements its own matrix math, one would expect functions for computing a determinant. Initially, these might be written in terms of explicit dimensions, with function names such as `det2` and `det3`. For convenience, a savvy coder might create a generalized `det` function, which determines the matrix size (in languages where this is available) and delegates to the appropriate dimension-aware function. All new code is written using this generalized `det` function, but old code remains with uses of dimension-specific versions. This is fine, so long as those functions also remain.

As time goes on, an even savvier team member improves the generalized `det` function. Instead of delegating to the dimension-specific functions, it now runs the proper summation across permutations of the elements. Seeing that `det2` and `det3` are no longer needed, he or she removes them. Tests will reveal any features whose code relies on the original versions, and these can be replaced with the generalized version.

Without tests, programmers begin to fear removing code. Setting aside the Frankensteinian notion of a programmer who has come to fear his or her own program and focusing instead on the resulting mess: Modules of code grow ever longer by collecting vestigial functions that no single person is certain are necessary, but neither is anyone certain that they are unnecessary. For how many months, or years, might they have been unnecessary? This extra code means that there is much more to hold in one's head when simply deciphering existing code, to say nothing of making large changes to it.

Any project that has functions, classes, or modules with names such as `writeResults-OldVersion` or `readPosition2` is crying out for refactoring, the likes of which is difficult in the absence of tests. In cases where these vestigial versions are found to still be necessary, testing them will help identify common functions or concepts that are ripe for abstraction. *Abstraction* is a term that programmers use to mean a kind of code generalization that increases flexibility and allows the programmer to work at higher conceptual levels.

17.2.4 Informing Code Usefulness

Writing automated tests for code is, at its heart, the simple act of writing more code that utilizes the code under test (the code which is being tested). If it seems difficult to construct a test for a particular module or function of code, then it is usually also difficult to reuse that module or function in other parts of the project. Thus, difficulty in writing a test is not simply a burden to grin and bear but is a sign that the code under test could be made simpler, for instance, deconstructing a function into several smaller ones, changing a data structure, or simply revisiting the assumptions of a module that groups together several functions that might not belong together.

This feedback about the simplicity of code is especially useful when a programmer can both code and test in short iterations, that is, alternating between writing code and writing tests often instead of coding for several hours (or days) and writing tests only after major milestones. This feedback is one of the advantages of test-driven development, the topic in Section 17.5.2.1.

Code that is difficult to understand is simply less useful as time passes from its moment of writing. Consider the idea of writing code in a manner so clever that it requires the upper limit of one's understanding to decipher. To write code this way is to make an assumption that everyone should find unnerving. It assumes that at all moments in the future (e.g., years later, during a thesis defense), the same programmer must be equally clever just to understand what he or she had once produced.

On the topic of complexity, an attitude shared by far too many researchers, scientists, and especially professional software engineers is the attitude of taking pride in producing complexity. This is natural, as in all technical fields, being capable of understanding complexity is held in high regard. It is key to realize, however, that domain complexity is not an excuse for code complexity! Being on the scientific cutting edge and doing novel research does not excuse shoddy, messy, or indecipherable code. People need to understand each other's code in order to have confidence in their work.

17.2.5 Tests as Documentation

Documentation is something that few software projects ever claim to have too much of. The opposite is more common: documentation exists, but it is sparse and possibly out of date.

Particularly when working as a team, automated tests serve as a form of documentation. Having code that demonstrates one or more uses of a function says to the code's reader,

"Here is how I, the author, intended for the function [illegible] to be used," by having code that uses that very function. This can cut through ambiguous or jargon-filled naming for functions or modules and help shed light on a function's purpose. Even better, tests can be executed to quickly verify that they are still accurate. Written documentation does not share that advantage. That said, tests do not replace documentation entirely, especially high-level overviews or architectural documentation.

17.3 The Mechanics of an Automated Test

17.3.1 The Four-Phase Pattern

There are several methods of enumerating the common steps of software tests. One of the oldest and most common explains testing in four phases: setup, exercise, verify, and teardown. These four steps are repeated for each test, regardless of its scope or scale. As will be discussed later, a single test may refer to testing just one aspect of one function or testing an entire program, and there are reasons to have tests in both categories.

17.3.1.1 Setup

A system under test is constructed. The phrase "system under test" is a fancy way of saying "the function, module, or program that this test cares about." This often refers to assembling or specifying the input data for which a programmer hopes to later verify produces a particular output or side effect. Examples of this include hard-coded values (numeric or otherwise), computationally creating input data, or retrieving known test data from an external source, such as the file system or a database.

In some cases, this step may be shared across tests to avoid prohibitively expensive computation time. If possible, keeping intermediate, precalculated input values is desirable to recomputing across multiple cases from some raw source data.

17.3.1.2 Exercise

In this step, the system under test is executed or invoked. Depending on the scale of the system under test and the goal of the test, this may range from a single function call to a sequence of a dozen or more.

17.3.1.3 Verify

The results of the exercise step are compared to an expected value. The notion of results falls into two categories: direct values returned by calling of functions and side effects on other parts of the system that are observable (e.g., data written to a file, database, or network connection). This notion of returned data versus side effects is discussed further in Section 17.5.1.

The act of verifying a test consists of one or more assertions, which cause the test to be marked as having *passed* or *failed*. In the case of a failed test, it is useful to provide a detailed message as to what specific expectation was not met. The level of detail may need to correspond to the scale of the system under test.

It is not common to provide detailed feedback for passing tests.

If the exercised code, or test code itself, causes an error, it is often treated as a separate outcome from that of an exercise that completed but whose result was incorrect. So tests can commonly be reported as one of three verifications: *passed*, *failed*, or *error*.

As with setup, any expected value used for comparison or assertion is most useful when it is accessible as directly as possible. Running calculations for comparison, risks introducing parts of the system under test into its own verification, invalidating its utility. Wherever possible, the storage of preprocessed expected values is desirable.

17.3.1.4 Teardown

The system is returned to its pre-setup state. Any records of test results or of computed values that need to be persisted should be written. In systems or languages with manual management of resources such as memory, file descriptors, database connections, network sockets, or the like, freeing these resources should occur during this phase.

17.3.2 Additional Testing Terminology

17.3.2.1 Test Case

When writing multiple tests for a specific section of code, for instance, to test behavior at the boundaries of valid input (also called an edge case), often, the entire set of these are referred to collectively as the tests, with each individual clause or subunit referenced as a test case to differentiate them.

17.3.2.2 Test Harness

A framework (flexible program library) that assists in setting up resources for tests and possibly provides advanced assertion capabilities not native to the language or environment is known as a test harness.

For instance, rather than having every single setup phase connect to an external resource (database, third-party service, logging system) before each of the hundreds of tests in a test suite, a harness will set this up once on behalf of all tests, which are then executed in the environment the harness provides. Most harnesses also help unify the output format of results. This is often a list of each test passing, failing, or having caused an error, as they occur. More advanced harnesses might produce an itemized final report saved somewhere, such as a shared web portal, for future reference by a team.

17.3.2.3 Test Fixture

Tests must be reproducible. Being able to rerun a test with the same input data is one important aspect of this. While a test harness provides common tools and procedures, *test fixtures* are data sources that one or more test case may operate on (use as input).

Fixtures work best when stored as static data (so-called "hard coded" values), but this is not always possible. However, whenever data are dynamically generated, it must be reproducible with ease. The most common example of this is utilizing a random number generator. When doing so, always use a seed value and provide some method of specifying the seed value for a test run. Seed values allow for the same sequence of random numbers to be reliably generated. Recording this seed value by including it with the test output will allow for reproducible successes and failures. A test suite with transient failures is simply not useful.

17.3.2.4 Test Coverage

A test or test case is said to "cover" a piece of code if it executes the code and makes assertions about its functionality. The term *test coverage* or sometimes *code coverage* is often used to discuss what percentage of a project's code is exercised by a full run of its tests. In practice, tools that help put a number on test coverage tend to be quite rudimentary, simply counting the lines of source code involved in execution and comparing that to total lines of code.

As an example of the shortcomings of this method, imagine a test suite for a project that implements basic arithmetic. A code coverage tool might count multiplication as being covered when a single test is run with input parameters of `1` and `1`, as in `multiply(1,1)`. Despite this test suite having covered the `multiply` function in terms of line counting, this metric could be fooled by implementing `multiply` in a way that always produced the value `1`. No sane programmer would be satisfied with that test alone, but it demonstrates that code coverage, while sometimes a useful heuristic, is not a true measure of the quality of a test suite.

17.4 Major Types of Automated Software Testing

With software having grown so integral to much of research, never mind the broader society, one might already be familiar with the phrases "stress testing" and "scalability testing" from the popular lexicon. However, these are less frequently occurring aspects of software testing, relative to the most common three or four: unit tests, functional tests, integration tests, and system tests.

This ambiguity of "three or four" arises because, for some programs or applications, there is little distinction to be made between functional and integration tests. As will be explained, unit tests are the smallest possible scope of verifying program behavior, and system tests are the most expansive and all-encompassing. The space in between is not always wide enough to discern two subcategories of tests, but as applications grow, there arises a difference between functional and integration tests.

Another note on ambiguity and terminology: software is a fairly young discipline, so some literature or online guides invert the scale of functional and integration tests such that integration tests are proposed as a finer granularity than functional tests, or use "functional test" in place of "system test" as the largest scale of test. Presented here, is the more common notion that the major types of tests, ordered from smallest scope to largest scope, are unit, followed by functional, then integration, and finally system tests.

17.4.1 Unit Tests

A unit test typically exercises a single function. The goal is to have the finest granularity possible. For object-oriented systems, the unit may be a class, especially if that class has a limited public interface. Many unit test enthusiasts will adhere to a rule of one assertion per test case, but this is not a strict requirement for a test case to be considered a unit test. It is nonetheless preferable to limit the number of assertions in each case and only group together those assertions that are truly related within the project's domain. Judicious division of assertions across multiple cases provides the most fine-grained, and thus most informative, feedback. This is particularly true when edge cases for a function or module are involved.

Another important aspect of well-written unit tests is isolation of the test cases from other units of code in the project. In high numbers, unit tests are the most powerful and useful tests to have, as they provide high levels of insight and confidence into the workings of each component of a project. Isolation is one key aspect of this power.

As a simple example, imagine a module that issues a greeting to persons in the language of their current location. The implemented module might have two additional functions:

- One that accepts a latitude and longitude and returns the local language
- A second that accepts a first name, last name, and language and returns a greeting to the named person in that language

greeter.py

```
# -*- coding: utf-8 -*-
def determineLanguage(country, state):
    if country == "United States":
        return "eng"
    elif country == "Germany" and state == "Bavaria":
        return "bar"
    elif country == "Germany":
        return "deu"
    else:
        return "eng"

def buildGreeting(firstName, lastName, language):
    if language == "eng":
        return "Hello " + firstName + "."
    elif language == "bar":
        return "Grüß Gott " + firstName + "."
    elif language == "deu":
        return "Guten Tag " + firstName + "."
    else:
        return "Hello " + firstName + ". Do you understand English?"
```

Unit tests for the Greeter module would aim to test all cases of both functions. Beginning with tests for determineLanguage:

greeter_unit_tests.py

```
# -*- coding: utf-8 -*-
import unittest
from greeter import *

class GreeterUnitTests(unittest.TestCase):

    def testIdentifiesEnglish(self):
        language = determineLanguage("United States", "Texas")
        self.assertEqual("eng", language)
    def testIdentifiesBavarian(self):
        language = determineLanguage("Germany", "Bavaria")
        self.assertEqual("bar", language)
```

```
    def testIdentifiesGerman(self):
        language = determineLanguage("Germany", "Brandenburg")
        self.assertEqual("deu", language)

    def testLanguageDefaultsToEnglish(self):
        language = determineLanguage("Chile", "Valparaíso")
        self.assertEqual("eng", language)

if __name__ == "__main__":
    unittest.main()
```

For those who wish to run these unit tests but are new to Python, they only need to run this command from a directory containing the above two files:

```
$ python greeter_unit_tests.py
```

A common mistake of programmers new to unit testing is to repeat function calls and assertions when testing a function that seems to build upon code used in a previous test case. This is tempting because it mirrors how normal program code might use these functions. For instance, they might leave in code that uses `determineLanguage` when writing a unit test for `buildGreeting`:

greeter_unit_tests.py

```
class GreeterUnitTests(unittest.TestCase):

    def testGermanGreeting(self):
        language = determineLanguage("Germany", "Berlin")
        self.assertEqual("deu", language)
        greeting = buildGreeting("Otto", "Valdez", language)
        self.assertEqual("Guten Tag Otto.", greeting)

    # Previous tests remain unchanged.
```

This is not ideal. It needlessly retests code that is already covered by `testIdentifies German`, duplicating work for no extra feedback on the code. Even worse, if a change to `determineLanguage` broke its identification of German, a test harness would report that `testGermanGreeting` failed, when the function and case it claims to test, namely, a German greeting, remains correct.

Instead, providing the expected result of `"deu"` from `determineLanuage` directly into `buildGreeting` is preferable for isolating the test of its behavior:

greeter_unit_tests.py

```
class GreeterUnitTests(unittest.TestCase):

    def testGermanGreeting(self):
        greeting = buildGreeting("Otto", "Valdez", "deu")
        self.assertEqual("Guten Tag Otto.", greeting)

    # Previous tests remain unchanged.
```

17.4.2 Functional Tests

Functional tests encompass multiple units, normally within the same module. These are expected to duplicate some code execution from unit tests and verify that the units work correctly when used in concert with each other. In many cases, this will simply be a number of calls to a handful of functions. In object-oriented languages, several classes might be tested for their ability to interoperate.

Using the previous greeter example, if one wanted to extend it to accept input in the form of geographic coordinates, he or she might do so by implementing these additional functions:

- A `determineAddress` function, which accepts a latitude and longitude pair and returns an address string. For simplicity in our example, this is assumed to be handled by an external resource.
- A `parseCountry` function and `parseStateOrProvince` function, both of which accept an address string and return a country and the proper next-tier political division, respectively, by counting comma-separated parts of the address.

greeter.py

```
# -*- coding: utf-8 -*-
# greeter module
import re

def determineAddress(lat, long):
    # Some external geographic database lookup.

def parseCountry(address):
    addressSections = re.split("\s*,\s*", address)
    indexOfCountry = len(addressSections) - 1
    return addressSections[indexOfCountry]

def parseStateOrProvince(address):
    addressSections = re.split("\s*,\s*", address)
    indexOfState = len(addressSections) - 2
    return addressSections[indexOfState]

def determineLanguage(country, state):
    # Unchanged from previous definition.

def buildGreeting(firstName, lastName, language):
    # Unchanged from previous definition.
```

While unit tests would be used to ensure that each of these new functions is operating properly in isolation, a functional test is used to ensure that the output of one can be used as the input of another in the manner the programmer intends:

greeter_functional_tests.py

```
# -*- coding: utf-8 -*-
import unittest
from greeter import *
```

```
class GreeterFunctionalTests(unittest.TestCase):

    def testGermanGreeting(self):
        address  = determineAddress(52.539, 13.433)
        country  = parseCountry(address)
        state    = parseStateOrProvince(address)
        language = determineLanguage(country, state)
        greeting = buildGreeting("Otto", "Valdez", language)
        self.assertEqual("Guten Tag Otto.", greeting)

    def testEnglishGreeting(self):
        address  = determineAddress(29.717, -95.401)
        country  = parseCountry(address)
        state    = parseStateOrProvince(address)
        language = determineLanguage(country, state)
        greeting = buildGreeting("Otto", "Valdez", language)
        self.assertEqual("Hello Otto.", greeting)

if __name__ == "__main__":
    unittest.main()
```

The scope of what a functional test covers can be somewhat flexible. When a set of three, four, or more functions are primarily intended to be used together, functional tests can be especially useful to maintain correctness even in the presence of passing unit tests.

If one wanted to introduce a new language code for Texas of `"tex"` as distinct from English spoken elsewhere, he or she might begin by adding it to `determineLanguage` and write a new unit test case for it:

greeter.py

```
def determineLanguage(country, state):
    if country == "United States" and state == "Texas":
        return "tex"
    elif country == "United States":
        return "eng"
    elif country == "Germany" and state == "Bavaria":
        return "bar"
    elif country == "Germany":
        return "deu"
    else:
       return "eng"

# Other functions unchanged from previous definition.
```

greeter_unit_tests.py

```
class GreeterUnitTests(unittest.TestCase):

    def testIdentifiesEnglish(self):
        language = determineLanguage("United States", "Washington")
        self.assertEqual("eng", language)
```

```
    def testIdentifiesTexan(self):
        language = determineLanguage("United States", "Texas")
        self.assertEqual("tex", language)

    # Other test cases unchanged from previous definition.
```

Now, there is a "tex" language option and GreeterUnitTests passes. Despite this, executing GreeterFunctionalTests will fail the testEnglishGreeting case, because (29.717, –95.401) happens to be in Texas, and without an entry for "tex" in buildGreeting, the input ultimately results in "Hello Otto. Do you understand English?" rather than simply "Hello Otto." or perhaps an enthusiastic "Howdy Otto!"

Saved by the presence of functional tests, the programmer recognizes his or her error and amends the buildGreeting function:

greeter.py

```
def determineLanguage(country, state):
    if country == "United States" and state == "Texas":
        return "tex"
    elif country == "United States":
        return "eng"
    elif country == "Germany" and state == "Bavaria":
        return "bar"
    elif country == "Germany":
        return "deu"
    else:
        return "eng"

def buildGreeting(firstName, lastName, language):
    if language == "tex":
        return "Howdy " + firstName + "!"
    elif language == "eng":
        return "Hello " + firstName + "."
    elif language == "bar":
        return "Grüß Gott " + firstName + "."
    elif language == "deu":
        return "Guten Tag " + firstName + "."
    else:
        return "Hello " + firstName + ". Do you understand English?"
```

And finally an update to the related functional test case:

greeter_functional_tests.py

```
class GreeterFunctionalTests(unittest.TestCase):

    def testTexanGreeting(self):
        address  = determineAddress(29.717, -95.401)
        country  = parseCountry(address)
        state    = parseStateOrProvince(address)
        language = determineLanguage(country, state)
        greeting = buildGreeting("Otto", "Valdez", language)
        self.assertEqual("Howdy Otto!", greeting)
```

It is not recommended to attempt to cover every possible combinatorial outcome of the inputs through functional tests. Even a half dozen functions coordinated would make the number of such tests prohibitively large. Instead, choosing one or two "happy path" cases and several prominent error cases will suffice to guard against major changes in the interoperability and coordination of the units under test.

If in the example, despite `GreeterFunctionalTests` not being comprehensive, if `determineLanguage` was altered to identify languages with simple number codes, this would be caught by `GreeterFunctionalTests` because the protocol of exchange between `determineLanguage` and `buildGreeting` would be altered.

A logical extension of this is that a project will tend to have fewer functional tests than unit tests, but that they cannot be relied on for comprehensive correctness.

17.4.3 Integration Tests

While functional tests may simply verify that several functions can cleanly hand off data between each other, integration tests can be thought of as a more human-scale assertion. These are especially relevant to a project with a graphical user interface (GUI) layer, which lends itself to having distinct user actions, or to projects that integrate with external resources or services, such as a database or third-party application programming interface (API).

Integration tests may also be useful as insight for monolithic projects whose many orthogonal functions are nonetheless grouped together in the same codebase. In this case, an integration test can be useful to someone outside the research team who is expected to frequently interact with the system or its code but who may not have full understanding of its construction at a low level.

Most useful of all is a property that integration tests share with functional tests: they help maintain a sense of what correctness means at a high level, allowing the details of implementation to vary as the researcher's needs or programming skills change. If a program is written to test a concept and a year later is rewritten such that the algorithm at its core changes (for performance reasons, to test a different model, or to simply clean up code for collaboration), then its functionality can be compared to the original version simply by running the existing integration tests.

A revealing example is image similarity. Whether or not any two images are similar can be calculated by many methods, each of varying quality when compared to how a human observer might judge the similarity of images. A very straightforward method is to simply measure the average color. This doesn't end up being a very useful program, especially for photographs, which tend to cluster around grays and browns when averaged. A refinement to the program might use k-means clustering to find the so-called dominant colors of an image and consider images similar if three out of four dominant colors are very close to each other.

At a high level, such a program might accept a list of images and return a list of mappings, specifying true or false for whether an image is similar to every other image in the list.

An example integration test for such a program might have two test cases:

1. Find that an image is similar to itself
2. For a set of three images, with two deemed similar, test that similarity is correctly determined

In Python, such an integration test might look like this:

```
import unittest
import image_similarity

class ImageSimilarityTests(unittest.TestCase):

      def test_sees_images_as_self_similar(self):
            images = ["jellybeans.jpg", "jellybeans.jpg"]
            expectation = {
                  "jellybeans.jpg": { "jellybeans.jpg": True },
                  "jellybeans.jpg": { "jellybeans.jpg": True }
            }
            results = ImageSimilarity(images).process()
            error_message = "Image was not found similar to itself"
            self.assertEqual(expectation, results, error_message)

      def test_similarity_of_three_sunsets(self):
            images = ["sunset1.jpg", "sunset2.jpg", "sunset3.jpg"]
            expectation = {
                  "sunset1.jpg": {
                        "sunset2.jpg"   : True,
                        "sunset3.jpg"   : False
                  },
                  "sunset2.jpg": {
                        "sunset1.jpg"   : True,
                        "sunset3.jpg"   : False
                  },
                  "sunset3.jpg": {
                        "sunset1.jpg"   : False,
                        "sunset2.jpg"   : False
                  }
            }
            results = ImageSimilarity(images).process()
            error_message = "{} and {} were not found to be similar"
            for filename in images:
                  for other_file in expectation[filename].keys():
                        self.assertEqual(
                           expectation[filename][other_file],
                           results[filename][other_file],
                           error_message.format(filename, other_file))
```

In the above code, even though the module and parent class of our tests is named `unittest`, this is only because that is the name of the Built-in Python test framework. What defines something as a unit test or not is the scope of the code it covers, rather than the name of the Built-in libraries it leverages.

Not shown here is the actual implementation of the `ImageSimilarity` class. This draws attention to an important point: the implementation is irrelevant to the test. This comes as a shock to the technically inclined, who tend to think in detail-oriented terms wherever possible. Instead, consider what the integration test is actually doing: asserting an expectation about the behavior of an entire, large program feature. While the mind naturally drifts to innards of the k-means similarity test, it is also asserting details about the program's ability to read files and about the result format.

In the result format presented in the test, in a comparison of all images to all images, the program might be duplicating work by comparing `sunset1.jpg` to `sunset2.jpg` and later comparing `sunset2.jpg` to `sunset1.jpg`. The researcher might choose to *memoize* this comparison (i.e., saving the results of computations to reduce duplicate work). Upon completing this performance optimization, the same integration test can be run to ensure that none of the changes altered the behavior of the actual similarity tests.

Thus, the implementation is free to vary, while the project's goals do not change. And while the project's goals do not change, the integration test does not need to either. Feedback from the integration test in fact assists in further refinement of the code.

It follows that if the project's expectations change significantly, the tests codifying those expectations might be expected to change. If at a later time, the researcher decided to implement the Shape Context algorithm as a measurement of similarity, this would likely have such a different result than color-based similarity that new test data and new expectations could be needed.

17.4.4 System Tests

Sometimes colloquially referred to as "end-to-end tests," system tests attempt a full run of all code under known conditions and assert only a final result. For a program with sufficiently narrow scope, integration tests and system tests may be one and the same. For larger projects, system tests are to integration tests as functional tests are to unit tests: they assert proper interoperability across ever-larger components, compared to the more focused test type.

In the case where final results need to be verified against the expectations of a theoretical model, any automation in that process almost certainly qualifies as a system test.

17.5 Knowing What, When, and How to Test

Testing is only useful insofar as it helps a project succeed. Like having too much documentation, rare is the project that has too many tests. But certainly, some tests end up being more valuable than others for a specific research project or for a certain phase in a project's lifespan.

17.5.1 Separating Function Concerns

If we categorize functions based on their responsibilities, they can fall into two broad categories:

- Query functions, whose purpose is to return a result, sometimes referenced as "data-in, data-out" functions (e.g., compute the Jacobian of a matrix, identify the country of a geographic coordinate). Tests for queries are the simpler of the two; they only need to make assertions about the result of the function call.

- Command functions, whose purpose is to cause a persistent side effect (or state change) rather than deliver a computed value. Tests for command functions will need to inspect values wherever the state change is stored in order to compare with an expected result. This may be an output file, a complex global data structure, an object's instance variable, or any other number of locations where program state is recorded.

Although there are sometimes performance reasons to do so, it is preferable to avoid writing functions that mix the responsibilities of being both a query and a command. The benefits of avoiding this are ease of code reuse and ease of testing.

And when testing is made easier, it is more likely to get done!

17.5.2 Testing Styles

There are a handful of methods for including tests in the day-to-day workflow of programming that have risen to popularity over the years. Two prominent examples are discussed here.

17.5.2.1 Test-Driven Development

An "inside–out" or "bottom–up" flavor of testing with a focus on unit tests goes by the name test-driven development, or simply TDD. Particular emphasis is placed on writing unit tests at the same time one writes the relevant functions, modules, or classes. A common workflow seen in TDD, and depicted in Figure 17.1, is known as "red, green, refactor," wherein a test for a function is written before the function code itself:

- *Red*: Initially, the test cases fail, as the function under test does not yet exist.
- *Green*: The programmer then writes the relevant function and runs the test any time he or she believes to have succeeded. The programmer knows that he or she has the correct code when the test cases finally pass.
- *Refactor*: With the code working, it can be cleaned up (and validated by keeping the tests passing) before moving on to writing another function.

17.5.2.2 Behavior-Driven Development

An "outside–in" or "top–down" style of testing, which begins with integration or functional tests, is known as behavior-driven development, or BDD. This style does not eschew unit tests but adds them when the need arises to codify finer points of logic. Often, the same red, green, refactor strategy is applied, but at a level that encompasses more code at once and where the jump from red to green (i.e., failing tests to passing tests) will take longer. A visualization of the difference between TDD and BDD is depicted in Figure 17.2.

One might wonder, why start at the integration or functional level and not even higher, such as with system tests? System tests, when used in a project large enough to set them apart from integration tests, are normally too high of a level to provide useful feedback during development, and integration tests tend to provide a more balanced place to start.

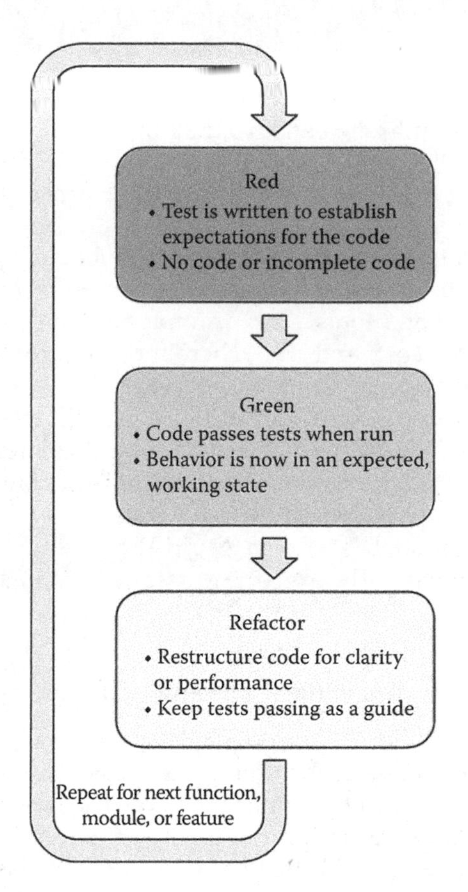

FIGURE 17.1
Red, green, and refactor workflow.

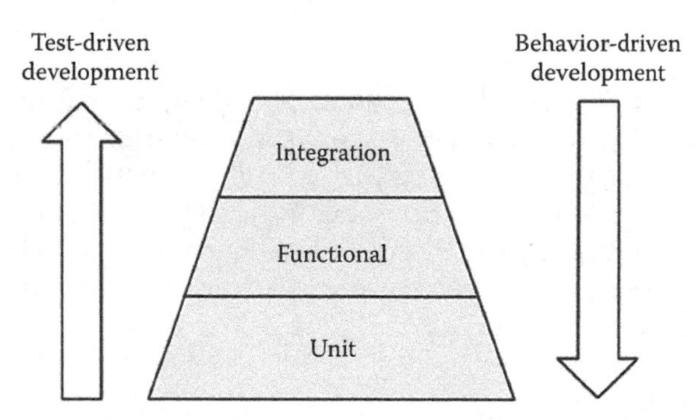

FIGURE 17.2
Test-driven versus behavior-driven development.

17.5.2.3 A Note on Testing Methodologies

Internet searches for TDD or BDD may bring up Internet arguments, with zealots debating the superiority of their favored style or other minute details thereof. Ignore the religious debates and focus on the benefits: getting feedback about one's code.

It is far more important to have tests that suit a project's needs than to slavishly implement someone else's ideals.

17.5.3 Considerations for Testing Parallel Code

Parallel code is as notoriously difficult to test as it is to write. Wherever possible, test a synchronous/serial version first to ensure that any errors are arising from the parallel execution and not incorrectness of the algorithm itself. One strategy to assist in this is to isolate the behavior of each concurrent thread or process from the mechanism of concurrency.

The simplest example of this is to have the primary behavior or calculation in one function, and another that orchestrates any threads, processes, or other parallel mechanism. Depending on environment and tools, even a simple solution such as a separate function may come at a performance cost, and thus, there may be a tradeoff between the ease of code organization and the needs of a running system.

If tests are written such that concurrent code is explicitly interleaved (e.g., to expose a race condition), there must be a method of reproducing this interleaving for the test to be useful. Determinism can be guaranteed via seeding the order or by using known test data, even if that data are in some part dynamically generated.

Aiding in this are tools such as Open Multi-Processing (OpenMP), discussed in Chapter 13, which allow explicitly executing parallel code as if it were occurring in serial.

17.5.4 Considerations for Testing Object-Oriented Code

In projects that heavily leverage the object-oriented features of a language, such as Python, it can be useful to treat a single class as the unit for unit tests. This should be employed, however, only when the class has a limited public interface. The larger a class becomes, the less sense it makes to treat it as a singular unit to test. Bear in mind the advice laid out previously: If the code feels unwieldy or difficult to test, then the class itself may be better off decomposed into smaller classes.

17.5.4.1 Testing a Class as a Unit

Referring to the program described in Section 17.4.3, imagine that the k-means functionality is implemented using four classes: `Pixel`, `ColorPoint`, and `Cluster`, being little more than data structures with a single `color_distance` method, and the `KMeans` class, which performs the bulk of the calculations, but with only one method that is not marked as internal: `results`. Calling this method will make use of all the others and return a result of the dominant colors for the given group of pixels. The code implementation below is followed by a discussion of testing it.

```
from random import sample
from math import sqrt

class Pixel(object):
    def __init__(self, x, y, r, g, b):
        self.x = x
        self.y = y
        self.r = float(r)
        self.g = float(g)
        self.b = float(b)
```

```
    def color_distance(self, other):
        return sqrt(
            ((self.r - other.r) ** 2) +
            ((self.g - other.g) ** 2) +
            ((self.b - other.b) ** 2)
        )

class ColorPoint(object):
    def __init__(self, r, g, b):
        self.r = float(r)
        self.g = float(g)
        self.b = float(b)

    def color_distance(self, other):
        return sqrt(
            ((self.r - other.r) ** 2) +
            ((self.g - other.g) ** 2) +
            ((self.b - other.b) ** 2)
        )

class Cluster(object):
    def __init__(self, pixels):
        self.pixels = pixels
        self.calculateCenter(self.pixels)
        self.r = self.center.r
        self.g = self.center.g
        self.b = self.center.b

    def calculateCenter(self, pixels):
        pixel_count = len(pixels)
        self.center = ColorPoint(
            sum([pixel.r for pixel in pixels]) / pixel_count,
            sum([pixel.g for pixel in pixels]) / pixel_count,
            sum([pixel.b for pixel in pixels]) / pixel_count
        )

    def color_distance(self, pixel):
        return self.center.color_distance(pixel)

    def __repr__(self):
        return "Cluster(rgb: ({r}, {g}, {b}),
                        total_pixels: {num_pix})".format(
                            r=self.r,
                            g=self.g,
                            b=self.b,
                            num_pix=len(self.pixels)
        )

class KMeans(object):
    def __init__(self, pixels, k,
                 iteration_limit=5,
                 convergence_limit=5.0,
                 initial_clusters=None):
```

```
        self.pixels = pixels
        self.iteration_limit = iteration_limit
        self.convergence_limit = convergence_limit
        self.active_clusters = self._determine_sample(initial_clusters, k)
        self.final_clusters = None
        self.computed = False

    def results(self):
        if not self.computed or self.final_clusters is None:
            self._compute()
        return self.final_clusters

    def _determine_sample(self, initial_clusters, k):
        if initial_clusters is not None and len(initial_clusters) is k:
            return initial_clusters
        return [Cluster([pixel]) for pixel in sample(self.pixels, k)]

    def _compute(self):
        current_iteration = 0
        prev_clusters = None
        while not self._beyond_limit(current_iteration, prev_clusters):
            prev_clusters = self.active_clusters
            self.active_clusters = [
                Cluster(new_pixel_group)
                for new_pixel_group
                in self._regroup_pixels(self.active_clusters)
            ]
            current_iteration += 1
        self.final_clusters = self.active_clusters
        self.computed = True

    def _regroup_pixels(self, clusters):
        proximity_mapping = {}
        for cluster in clusters:
            proximity_mapping[cluster] = []
        for pixel in self.pixels:
            nearest_to_pixel = reduce(
                self._nearest, clusters, (pixel, None, float("Inf"))
            )[1]
            proximity_mapping[nearest_to_pixel].append(pixel)
        return proximity_mapping.values()

    def _nearest(self, (pixel, current_nearest, current_distance),
    cluster):
        this_distance = cluster.color_distance(pixel)
        if this_distance < current_distance:
            return (pixel, cluster, this_distance)
        else:
            return (pixel, current_nearest, current_distance)

    def _beyond_limit(self, current_iteration, prev_clusters):
        return (self._iteration_done(current_iteration) or
                self._converged(prev_clusters))
```

```
    def _iteration_done(self, current_iteration):
        return current_iteration > self.iteration_limit

    def _converged(self, prev_clusters):
        if prev_clusters is None:
            return False
        return all([
            prev.color_distance(current) < self.convergence_limit
            for prev, current
            in zip(prev_clusters, self.active_clusters)
        ])
```

In Python, methods prefixed with an underscore are not intended for use outside the class itself. This is similar to `private` methods in C++ or Java. So although `KMeans` has eight nonconstructor methods, actually using the class boils down to simply calling the `results` method.

In the case where internal methods of a class are tightly coordinating to accomplish some goal, some of the methods may not be intended to be used, or tested, separately. The first goal of testing a class as if it were a unit is to test all of the public methods, and here, there is only one. Thus, one can be certain of this class' correct behavior with only tests that instantiate new `KMeans` objects and then call `results` on them.

But is that correct? The beginning of the k-means algorithm takes a random set of `k` pixels to start with, and tests that cannot be reproduced are not useful. In pursuit of this, the constructor has an optional keyword argument named `initial_clusters`, which allows for reproducible calculations by overriding this random selection.

A final note on private or internal methods. Making methods private may appear pointless to some new programmers. After all, the source is right in front of them, and there is nothing private about that! However, from the perspective of code organization, the use of private methods is to separate the stable, public interface from the gory details that the class' author expects might change at any time or in any way. So, while it is considered poor form to wildly alter the public interface of a class just to optimize some internal details, when it comes to private or internal methods, all bets are off.

In the case of `KMeans`, the actual clustering is performed by the `_regroup_pixels` method, which is a simple query method, not relying on any state from the rest of the class. When working in a TDD style, if it would help the programmer feel more confident in his or her code, it might be appropriate to test this method (even though it is internal) and throw out those tests when the class is passing tests of its public methods. By definition, those public methods should be using the private method, and so tests for the private method are redundant in all but a few rare instances.

17.5.4.2 Inheritance and Testing

If your project is using inheritance, it is always worthwhile to ensure that all of the subclasses are truly acting as subsets of their parent class' behavior. In addition to testing any subclass-specific refinements to the code's behavior, it is recommended to run all subclasses through any tests intended for the parent class. Further reading on this idea can be found by exploring the Liskov Substitution Principle, which simply codifies this recommendation in formal mathematical terms.

17.5.4.3 Test Doubles

In systems that need to utilize many objects with interrelated references to other objects during execution (also called having a large object graph), constructing these for each test case, especially in unit tests, can be slow and difficult to maintain as the project changes.

For instance, to test a `Car` class in a simulation of a car, one might require building an `Engine` object, which requires many `SparkPlug` objects, a `CrankShaft` object, and so on, all which require their own subcomponents. Unit tests for the `Car` class are then implicitly also testing many of these other classes, making them more like functional tests than unit tests.

To alleviate this, and truly test only an instance of the `Car` class, tools known as test doubles may be employed. There are several tools that fall under this category, the most common being mock objects (often simply called "mocks"), and the core idea is to provide a fake object of some sort that gives known, dummy responses. Many test frameworks for object-oriented languages provide a mock functionality, and some will even provide tools to validate that a mock was used properly (e.g., by checking that certain methods on the object were called).

17.6 Concrete Uses of Testing

17.6.1 Understanding and Modifying Legacy Code

When given an existing project, one of the first steps to understanding the code is to run any existing tests. Reading the code for the tests that pass can be a very useful step in understanding how the code was intended to be used.

If there are no tests, before you modify any program behavior, it is useful to execute the code with several different sets of input and capture these as test cases. These are often called characterization tests, in that they help characterize the state of the program at a known point, without necessarily asserting that its behavior is correct or desirable. If the legacy system is known to be functioning correctly, these tests will help guide any changes. If the system is not known to be functioning correctly, there is at least a baseline to know what functionality has been affected by any changes the programmer makes.

17.6.2 A New Project with Many Unknowns

In a new project that will require significant exploration (in terms of code, rather than science), or any situation where requirements may be unclear or rapidly evolving, it often comes to pass that entire modules of code are thrown out and rewritten. This exploratory style of coding is called a code spike or simply a spike, and its goal is to learn more about the problem or domain, rather than to produce a specific, stable implementation.

In projects or project phases employing a spike style, an excessive focus on unit testing is not always as helpful as it would be in more stable projects. Loose integration or system tests can help the team decide when milestones have been reached. Unit tests can and should still be implemented but can be delayed until requirements stabilize.

17.6.3 A Project with Certainty in Its Goals

This applies to both adolescent projects, whose behavior is well established, and to new projects that have high certainty in their requirements. As with the previous, integration and system tests still help track progress toward milestones. Unlike projects with more uncertainty, any expectations or details that are known at the time of writing code should be captured in unit tests.

17.7 Realities of Testing in a Project

Even a reader who is excited by the confidence, simplicity, and clarity that an automated test suite provides may still have lingering questions about the feasibility of the extra work involved. In fact, a common criticism from undisciplined programmers tends to be that tests slow down development because they require writing twice as much code.

17.7.1 Does Testing Take a Lot of Time?

Writing software tests is a discipline. It is a muscle that must be developed and, like any muscle, becomes quicker and feels more natural over time. More experienced programmers and software engineers tend to be among the most adamant about testing. This is no coincidence. Testing often pays off in the short run and always pays off in the long run.

The type of tests chosen for a project, the amount, and what code is targeted by tests lie on a spectrum of usefulness. The notion of "code coverage" introduced in Section 17.3.2.4 does not have a direct correlation with software reliability. As discussed in Section 17.6, some projects gain more from a loose set of integration or functional tests, while other projects benefit greatly from rigorous unit testing of any significant function in the system.

17.7.2 How Many Tests Are Enough?

The sky is the limit, but it is important to consider what is useful to the project. Diligently keeping code coverage at 100% can feel rewarding (almost like a game), but major projects can be certain of functionality at 85% or 95% coverage. Avoid tests that are actually testing the environment, the language itself, or any third-party libraries (especially when those libraries have their own test suites).

For those wishing to push the limit on software quality, two advanced tools are introduced in Sections 17.7.2.1 and 17.7.2.2.

17.7.2.1 Mutation Testing

Mutation tests work by analyzing a project's source code; changing a single constant, symbol, or comparison operator; and then running the full test suite. If all tests pass despite the code having been "mutated," there must be a case that the tests don't cover. For instance, a mutation test may change a < comparison to a <=, causing an iteration to proceed longer than it ought. In this way, mutation testing is like a test for a project's test suite.

17.7.2.2 Property/Generative Testing

Like mutation tests, *property tests* test a project's test suite and often operate on entire types or ranges of data. For example, a property test for code that operates on n-grams might run a function on all possible combinations of character data. The common use of property testing is to identify overlooked edge cases, which can then be brought back into the project's unit tests as a specific test case. In some software communities, property testing is referred to as generative testing.

Since both mutation tests and property tests require multiple runs of your test suite, they can be computationally prohibitive and are not appropriate for all projects. However, selective use in mature projects can be very powerful.

17.7.3 Convincing Your Team

Often, the situation arises where one member of a group or team is the first to see the light of testing and wishes to spread the good news in hopes of saving everyone time and improving software quality for all. It is common to be met with apathy or dismissive attitudes, but the advocate need not lose hope!

Setting a good example with one's own code and tests is always the first step. After that, providing the team with an explanation or guide to the chosen harness or framework can do much to alleviate any fear of the unknown that might have taken hold.

It is common for people who have coded for a long time without testing to view the process as a burden, undertaken after the "real work" is done, as if to fulfill some bureaucratic checkbox. For these types, evidence that styles such as TDD or BDD can save time even in the short run, tends to be most persuasive.

If all else fails, writing high-level functional or integration tests for the project can help ease the team into a notion of automated testing. It is an unfair burden to have to write unit tests for everyone else's code, but after one's integration tests save the project from a few bugs (often some humble, overlooked edge cases), the argument for better testing tends to make itself.

Glossary

Abstraction: A method of delegating or concealing details of code's implementation so that the programmer can work at a higher, or more general, conceptual level while still benefiting from guarantees about specific functionality.

Behavior-Driven Development (BDD): A software development methodology that seeks to drive code functionality and software design toward project goals by defining tests that make broader statements about application or domain logic. Compared with test-driven development, BDD tends to be seen as "top–down" in its approach.

Codebase: The set of all source code in a project.

Code Spike: A phase of exploratory coding whose intent is to learn more about a domain or problem, rather than to produce stable, clean code. Code produced from a spike is often heavily in need of refactoring or perhaps thrown out and used only as a reference or rough draft for the actual implementation.

Code Under Test: The code that is being executed by a given test case and whose correctness is the subject of the said test. Often used in the context of unit tests or functional tests.

Command Function: Functions whose purpose is to cause a persistent side effect or observable state change. One of two broad categories of functions, from the perspective of testing strategies.

Edge Case: A situation or behavior that occurs at the boundary of the acceptable parameters for some code or system. Often used to draw attention to potential sources of errors or oversights in program logic.

Functional Test: A test that encompasses multiple units, verifying that they interoperate as the programmer intended.

Integration Test: A test that verifies the correctness of code at a higher conceptual level, often describing an entire feature or user action that might involve dozens of functions, classes, or even external resources. Integration tests maintain human-scale assertions about the program.

Liskov Substitution Principle (LSP): A mathematical description of strong behavioral subtyping in object-oriented languages. It says that an instance of a subclass should be able to stand in for any instance of its superclass (in test cases, for example). Adhering to the LSP contributes to the clarity of a program's design.

Memoization: Storing the result of a particular computation, especially computationally expensive operations, for reuse rather than recalculation.

Mock Object: A type of test double that provides known, "dummy data" responses instead of executing the code of the object that it stands in for.

Mutation Testing: Testing that begins by mutating a project's source code and then runs the entire test suite, expecting a failure. If no failures occur, this indicates a weakness or omission in the test suite.

Object Graph: In object-oriented languages, the execution environment will come to consist of many objects that contain references to each other in a graph structure.

Property Testing: A type of testing that runs a series of integration-like tests, but across programmatic ranges of input data in hopes of identifying overlooked edge cases or other problematic input.

Query Function: Functions that only seek to return a result for their input parameters. One of two broad categories of functions, from the perspective of testing strategies.

Refactoring: A reorganization of code that does not alter its behavior, often with a goal of increased clarity. Sometimes called "cleaning up" the code. Identifying and generalizing new abstractions are common tasks in the process of refactoring.

Seed Value: In tools such as random number generators, seed values are inputs to the initial algorithm that allow for the same sequence of random numbers to be reliably generated each time the same seed is provided.

System Test: A test that attempts a full run of all major codes under known conditions, asserting only a final result. Sometimes called "end-to-end tests." For small projects, system tests may be indistinguishable from integration tests.

System Under Test: Related to code under test, but describing a larger unit, often at a scale of an entire feature or application. It would be unusual to call a single function a "system under test" in the context of a unit test.

Test Case: The individual clause that describes a specific assertion about the code (or system) under test.

Test Double: In object-oriented languages, test doubles help isolate the part of the system currently under test from other parts upon which it may depend, but whose detailed behavior is unrelated to the particular test.

Test-Driven Development (TDD): A software development methodology that emphasizes short cycles of feedback between unit tests and the code being written in order to guide code functionality and the software's design toward the project's goals. When compared to behavior-driven development, TDD can be thought of as a "bottom–up" approach.

Test Fixture: Data sources, usually static, that one or more test cases may use as input. When dynamic, there must be a manner of seeding any random element or reselecting the same dynamic values, in order to reproduce any test failures.

Test Harness: A framework (or program library) that assists in one or more of the following: setting up resources for tests, providing assertion capabilities, or providing the presentation of test results, from simple console output to more advanced reports.

Unit Test: The most granular test type, typically exercising a single function.

18

Validation of Computational Models and Codes

Christopher R. Iacovella, Christoph Klein, Janos Sallai, and Ahmed E. Ismail

CONTENTS

18.1 Introduction

Validation is an essential process that should be undertaken before any computational model is used for "production" purposes. In general, validation helps ensure that the *inputs* to software are accurate, to the extent that they produce the desired *output* behavior. While validation has many parallels to verification, verification ensures that the components of the underlying software are correct, whereas validation focuses on ensuring the phenomenological correctness of a model—for instance, by testing that a set of parameters given to a simulation code produce the correct thermodynamic outputs. Thus, one can consider software to be somewhat decoupled from the model. For example, a programmer implementing a database can verify the core functions without needing to validate the specific data that will be stored in the database. This distinction can be summarized as follows:

- Verification: Are we building the system *right*? Is the code correct?
- Validation: Are we modeling the *right* system? Did we give the code the correct information?

Clearly, verification and validation are linked, as incorrectly programmed software would likely cause a model to fail validation. However, even the best codes can yield problematic results if we provide improper inputs. A key aspect of validation is the process of making sure we avoid the "garbage in, garbage out" phenomenon. A classic example of this is the Mars Climate Orbiter mission of 1998. The Climate Orbiter was designed to record information about the climate and atmosphere of Mars as part of a hunt for water resources on the surface and underground. However, because of an input/output error in which a code reported output in "traditional" units, whereas downstream codes expected the results to be in SI units, significant errors in the trajectory of the probe were introduced over time. The consequence of these collected errors was the likely destruction of the $125 million probe in the Martian atmosphere and total mission failure [1,2]. While most scientific computing applications will be on a much smaller, less costly scale than space exploration, this in no way trivializes the importance of validation.

18.2 What Is a Model?

In the most general sense, a model consists of a set of rules designed to mimic the key features of a system, phenomenon, or process. The rules in a model may be simple or complex, numerous or few, all depending on the application and what behaviors the model aims to capture. In the realm of scientific computing, we typically focus on *mathematical* models, whose rules are codified as equations and associated fitting parameters.

As a familiar example, consider modeling the process of a ball falling from rest. The simplest model would involve codifying the set of rules that describe the motion of the ball due to gravity (i.e., the kinematic equations). This model could be made more complex, and presumably more accurate, by adding rules that account for air resistance or the inelastic collision of the ball with the ground or by further taking into account the rotation of the ball over time. The level of detail required of the model depends on which properties or behavior we are ultimately interested in measuring and how accurately we intend to reproduce them. For example, a model that takes into account air resistance would likely be unnecessary if we simply want to reproduce the behavior of a ball falling a short distance, where such effects are negligible. Note that while the kinematic equations are the essential component of this model, the model does not necessarily provide the algorithm for solving these equations. A standard equation solver, such as Mathematica or MATLAB, is capable of mathematically evaluating the model irrespective of which phenomena the equations are actually modeling. Again, the model and software can typically be decoupled, allowing validation and verification to be performed independently.

In scientific computing, models can be developed from a variety of sources, including the following:

- Knowledge of the underlying physics of a problem
- Low-level first-principles calculations
- Other more complex or costly models
- Experimental measurements or observations
- Some combination thereof

For example, first-principle calculations can be performed to capture the energy required to form a bond between a gold atom and a sulfur atom; these measurements can then be used to develop a much simpler and computationally cheaper model [3]. In such an example, model validation would include both how accurately the model captures the first-principles calculations and how accurately the model reproduces experimental behavior, such as enthalpy or entropy changes. The specific metrics used to validate the output of the model will of course depend on the system or process that is being modeled.

Figure 18.1 shows a conceptualization of how mathematical models and experiment are often intertwined. Experiments, or more sophisticated calculations, yield hypotheses that

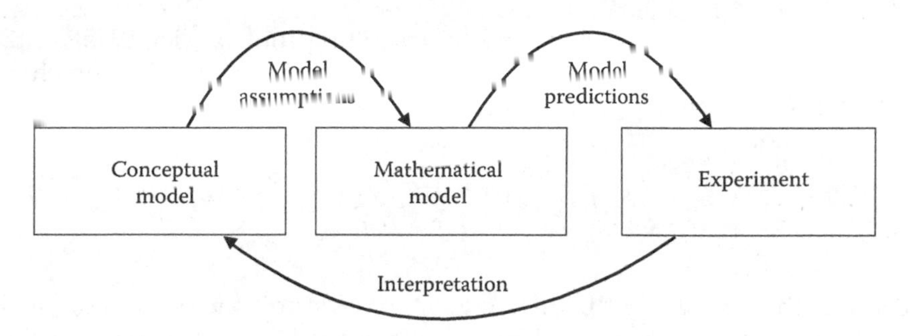

FIGURE 18.1
Example of a conceptualization of how models are developed and refined. An experiment typically yields data, which can be structured into conceptual models that propose a mechanism, from which mathematical models can be developed. Mathematical models can then be used to predict behaviors, which can be further tested in experiment. The entire process can then be repeated to refine the behavior.

enable the development of conceptual models (i.e., mechanisms are proposed based on the data). Mathematical models are then developed to test the conceptual models and yield new results, insight, and refinement of the proposed mechanisms. These refinements are used to design new experiments and test new hypotheses, and the process is then repeated. Different steps involve a varying degree of validation. If the mathematical model is unable to reproduce the behavior seen in the experiment (or a more expensive calculation), the model would require refinement to pass validation. Note that the individual steps in this process may not be tightly coupled to one another (i.e., performed in collaboration) and instead result from indirect communication through published datasets and literature.

Scientific models certainly exist on length scales beyond atoms and molecules. Examples are diverse, including mathematical models developed to capture the impact of signaling hormones on the bone remodeling process [4], galaxy formation [5], traffic patterns [6], and the manufacturing processes of composite materials [7]. Given the diversity of models in scientific computing, it is not possible to define a single list of validation steps that will work for all models. Instead, we will focus on providing several general validation schemes that highlight the thought processes that should be considered when validating a specific model. To provide a more tangible basis, these schemes will be discussed in the context of molecular models as a prototypical example of a scientific model. As such, we will first provide a basic overview of molecular models.

18.3 Molecular Models

When coupled with molecular simulation tools, molecular models (also called "force fields") have enabled researchers to gain understanding of atomistic- and molecular-level interactions and mechanisms that underlie natural and manmade materials and processes. Such models typically consist of a set of parameters and equations that describe how atoms (or more generally, particles) interact with each other. While molecular models typically include a large number of equations and parameters meant to capture both the nonbonded (van der Waals and electrostatics) and bonded (bonds, angles, and dihedrals) interactions, for simplicity, we will focus only on van der Waals interactions between atoms (Refs. [8,9] provide a very detailed overview of molecular modeling and simulation).

The most common way of mathematically describing the van der Waals interactions between particles is the 12-6 Lennard-Jones potential in Equation 18.1, which describes the pair interaction potential, U, between two particles as a function of their separation r:

$$U(r) = 4\epsilon\left[\left(\frac{\sigma}{r}\right)^{12} - \left(\frac{\sigma}{r}\right)^{6}\right], \tag{18.1}$$

where ϵ controls the strength of the interaction and σ controls the particle diameter; these two adjustable parameters vary based upon the chemical species being considered. The Lennard-Jones form is only one of many possible interaction potentials that can be used to describe the nonbonded interactions between atoms. The Morse potential shown in Equation 18.2 is another description, where D_0 represents the characteristic energy, r_0 sets the location of the potential minimum, and α adjusts the shape of the curve.

$$U(r) = D_0\left[e^{-2\alpha(r-r_0)} - 2e^{-\alpha(r-r_0)}\right] \tag{18.2}$$

Simplified interactions, such as the square-well potential in Equation 18.3, can also be used to capture the van der Waals interactions, where systems have three adjustable parameters: ϵ, which sets the interaction strength; σ, which sets the particle diameter; and λ, which sets the range of the interaction.

$$U(r) = \begin{cases} \infty, & r \le \sigma \\ \epsilon, & \sigma > r \le \lambda\sigma \\ 0, & r > \lambda\sigma \end{cases} \tag{18.3}$$

In the context of this chapter, the scientific details of these models are relatively unimportant. The key takeaway messages, however, are the following:

- There are often multiple ways of mathematically describing the same phenomena.
- Models typically have many fitting parameters that can be tuned to produce specific behavior.
- Which model to use depends on which behaviors we are trying to capture and the computational efficiency required.

In general, defining a molecular model in a simulation software package requires a user to do the following:

1. Define which pair potential to use. Note, users typically do not have to mathematically define this function; instead, the form of the function is generally specified by the developers of the forcefield. The end user needs to ensure only that the mathematical form is supported by the simulation tool.
2. Define all parameters required by the pair potential, such as σ and ϵ for the Lennard-Jones potential.
3. Define the initial position and type of all atoms in the system (as well as any bonds, angles, etc., required for more complex models).

Given that in most simulation software, functional forms of pair interactions are "Built-in," it will be assumed in this chapter that such simulation software has gone through proper verification, to ensure that the functional form and other related simulation schemes are implemented correctly. Additionally, it is important to note that initialization of the system configuration (i.e., defining the positions of the atoms and types) is often done using *ad hoc* scripts written by users and independent of the simulation software. Thus, in the case of molecular models, validation of the model also implicitly requires verification of the software used to generate the model, as will be discussed below.

18.4 Validation

As previously discussed, a key goal of validation is to avoid the "garbage in, garbage out" phenomenon. In this chapter, we make the assumption that the underlying software has been correctly implemented (verified) and all sources of error reside in the model itself. Note that while numerical or round-off error may also cause a model to fail validation, such factors will not be considered here as this topic is discussed in Chapter 2. Let's consider two main causes for the failure of a model:

1. Input parameters or equations passed to the code are not as expected. (Did we give the code the correct information?)
2. The parameters and equations are passed to the code correctly, but the model itself does a poor job of capturing the underlying phenomena. (Are we modeling the right system?)

18.4.1 Validation of Model Inputs

Let's first consider the process of validating that we have defined the model inputs correctly for a given piece of software. At the most basic level, validation of a model involves manually checking, and often rechecking, the associated input parameters for accuracy. In this regard, several pertinent questions must be considered.

Are the units correct? Are there any typographic errors?

For models or parts of a model with only a few inputs, these checks are fast and manageable. For example, a simple molecular model will require the definition of a few dozen input parameters for the various bonded and nonbonded interactions, which can be manually validated with minimal time commitment.

Is the formatting correct?

Beyond validating the numerical values, it is important to take note of the expected formatting of the parameters. Many pieces of software are inflexible and require values to be defined in a specific order, often with little context. For example in Listing 18.1, consider the script commands used to define the ϵ and σ parameters for the Lennard-Jones potential, in the syntax for both large-scale atomic/molecular massively parallel simulator (LAMMPS) [10] and highly optimized object-oriented molecular dynamics (HOOMD)-Blue [11] simulation engines.

LISTING 18.1 Definition of Lennard-Jones Parameters

```
#(LAMMPS)
pair_coeff 1 1 0.066 3.5
#(HOOMD - Blue)
lj. pair_coeff.set ('1 ' , '1', epsilon=0.066, sigma=3.5)
```

Clearly, LAMMPS provides less context in terms of labeling what each numerical value represents and, as a result, also requires all parameters to be entered in a specific order. This increases the likelihood of an error and, without clear labels, makes it more difficult to spot an error (for instance, if the numerical values are correctly defined, but in the wrong order).

Inputs to software may be even more restrictive, where some codes expect values not only in a specific order but also at fixed character locations, known as "fixed-width" formats. For example, if a code expects fixed-width data, the two lines in Listing 18.2 are *not* equivalent, even though they are mathematically identical. While the first value may be read correctly (i.e., 0.066) for both lines, the second value in the top line might be incorrectly read by the software as .5 rather than 3.5, as the 3 exists at a character location where the code expects a space. This could lead to significant errors that are difficult to detect, as the difference between 3.5 and 0.5 might not lead to an obvious runtime error (as discussed in Section 18.4.2).

LISTING 18.2 These Lines Are Not Identical for Fixed-Width Data Schemes

```
0.066 3.5
0.0660 3.5000
```

As the number of needed input parameters exceeds what is practical to check by hand, different strategies must be adopted. For example, as part of defining a model for a molecular simulation code, thousands of lines of input data are typically required to define the positions, types of atoms, which particles are bonded, and so on. Often, such input files are generated using *ad hoc* scripts written by the user that have not gone through any formal verification given their limited scope; unit tests or other common techniques are frequently not employed. As such, "verification" of the scripts often falls into the category of validation. As smaller files are easier to manually validate, it is often useful to generate models incrementally. For example, if we write a script to generate the position of atoms or molecules in a box, a first test to ensure that the code correctly generates the parameters would be to use a small number of atoms or molecules that can be manually validated with minimal effort. With confidence that a small number of atoms or molecules (e.g., 1–10) can be generated without error, one would have more confidence that a larger system, for which manual validation would not be feasible, would also be generated without error. While it is impractical to check each line of a large output file, a cursory visual scan through a large input file can help to identify errors, through the consideration of several questions.

Does the order of magnitude of the values in the file make sense?

For example, if the values are meant to define the positions of atoms in a box with edge length 100, we would not expect to find positions with values on the order of 1000 or more. Additionally, for molecular models, system configurations can be visualized using tools such as Visual Molecular Dynamics (VMD) [12], which can provide considerable insight into whether the system structure is as expected, such as visually determining whether all atoms exist within the confines of the simulation box.

Does the file feature many values that look to be "default" values?

For example, if a file prominently includes values that are often associated with undefined parameters, such as not a number (NaN), 0.000000, or numbers that resemble memory addresses (such as 1470617776 or a hex code), this may indicate an error in the script used to generate the model input file, typically associated with failing to assign a numerical value to a variable. While finding "default" values may certainly indicate an error, and the model would fail validation, the opposite—that is, not observing such values—does not necessarily imply there are no errors.

Automated checking of large files.

In specific cases, comprehensive, automated tools exist for validation of input files. For example, users of the GROningen MAchine for Chemical Simulations (GROMACS) [13] molecular dynamics package must first run their simulation inputs through a preprocessing tool called grompp. Among other things, grompp performs myriad validation checks on the provided input files. Many nonsensical user selections that would unequivocally result in incorrect outputs instead yield descriptive error messages. Dubious or nonstandard user selections may result in warnings such as tips to increase simulation performance. Similar approaches can be taken to automate many of the tasks described above, for example, checking input files for particles that exist outside the simulation box, ensuring that values are of the appropriate order, or providing warnings for values that may be problematic.

To help avoid issues with model definition and to make it easier to perform validation, best practices can be put into place to reduce ambiguity. For example, it is common that software allows users to add comments to the input scripts used to define the model. For example, to reduce ambiguity, one might include comments in a LAMMPS script, as shown in Listing 18.3, that describe the order of parameters, define the expected units, and/or list the source of parameters and what they are meant to describe.

LISTING 18.3 Examples of Providing Context and Documentation via Comments

```
#typeA typeB epsilon(kcal/mol) sigma (angstrom)
pair_coeff 1 1 0.066 3.5
#interactions for C-C from OPLS-AA
pair_coeff 1 1 0.066 3.5
```

Since *ad hoc* codes are frequently used to generate model inputs, care can be taken to write these codes in ways that minimize errors. For example, we can take advantage of libraries that allow variables to have units associated with them (such as the Python units package), which can help to ensure consistent and accurate unit conversion. The use of verification schemes can also help to avoid errors in models; in particular, the use of assert statements, and similar checks, can be used to avoid issues related to "undefined" parameters. For example, if we wish to model a system that contains 10,000 particles, we can include a simple check to ensure that 10,000 atoms have been defined. This is shown in pseudocode in Listing 18.4. Assert statements can also be used to ensure that files read by scripts exist in the location specified or that the file contains all the data expected (e.g., number of particles), as failure to properly load data into the script can result in undefined outputs in the model files the script generates.

LISTING 18.4 Pseudocode for Validation of Scripts

```
int N_total = 10000;
positionVector xyz;
file myInputFile;
...
assert(myInputFile.good () );
...
if(xyz.size () ! = N_total)
{
      print 'Error: system size. '
      print 'Found:', xyz.size ()
      print 'Expected:', N_total
}
...
assert(xyz.size () == N_total);
```

18.4.2 Sanity Checks

Before putting forth the effort to validate if a model produces the "correct" behavior, it is often faster and easier to first determine if it is incorrect. For example, as previously mentioned with regard to input parameters, if a set of values is meant to define the positions of atoms in a box with edge length 100, we would not expect to find positions with values on the order of 1000 or more. If values do exceed the box dimensions, we could immediately determine the model is not implemented correctly. Again, even if all particles exist within the confines of the box, that does not necessarily imply that the model is correct. Loosely speaking, this type of examination is considered a "sanity check."

In the realm of molecular models, other sanity tests could include examination of the pressure or energy of the initial configuration. If the potential energy or pressure is unphysically high (e.g., a pressure of ~10,000 atm for a simple liquid rather than ~1 atm), this may indicate significant overlap of particles and failure of either the scheme to initialize atoms/molecules or an incorrect definition of the σ parameter for the Lennard-Jones potential (recall, σ defines the size of the atom). This of course requires knowledge of the specific system being modeled, such that one understands what are reasonable values. As a further example of this, consider implementing a Lennard-Jones model of atoms at a liquid state. Table 18.1 reports the potential energy and pressure of the initial configuration of two systems at matching number densities, which differ only by the σ parameters for the Lennard-Jones interaction. Just from these properties, someone with knowledge of the system would immediately deduce an error in the system for $\sigma = 2.5$, given that the potential energy and pressure are atypical of a simple fluid system. While this does not indicate that the system with $\sigma = 1.2$ is an accurate model, it does pass the sanity check and indicates that $\sigma = 2.5$ is certainly an incorrect parameter.

Similarly, a sanity check can involve making sure that the outputs, such as thermodynamic quantities like pressure and potential energy, are comparable for similar systems.

TABLE 18.1

Potential Energy and Pressure for Two Otherwise Identical Systems with Differing σ Values for the Lennard-Jones Potential

σ	Potential Energy	Pressure
1.2	–4.0793157	0.3109708
2.5	28,840.844	46,701.566

That is, we would expect molecules such as decane ($C_{10}H_{22}$) and dodecane ($C_{12}H_{26}$) to produce similar outputs, given that they are both hydrocarbons that differ only by two carbon groups. However, such similarity would not be expected for dodecane and propane (C_3H_8), given the significant difference in carbon backbone length.

Ensuring that the system has the correct phase or structure is another straightforward sanity check. For example, the radial distribution function (RDF) is a typical way of identifying the phase of a system, where gases demonstrate RDFs that rapidly decay to unity and liquids tend to exhibit multiple peaks as a function of separation. Figure 18.2 plots the RDF of two systems with identical densities but with different ϵ parameters for the Lennard-Jones potential. One system predicts a gas state ($\epsilon = 0.1$) as compared to the liquid-like structure ($\epsilon = 1.0$), as shown in Figure 18.2. Thus, clearly, a model with $\epsilon = 0.1$ would fail the sanity check if we anticipate a liquid, and $\epsilon = 1.0$ would pass.

Visualization of the system may demonstrate obvious failures as well, such as particles outside the box, or a small subset of particles that appear to be moving much faster, or much slower, than others. Similarly, if a molecule is known to be very rigid but appears to

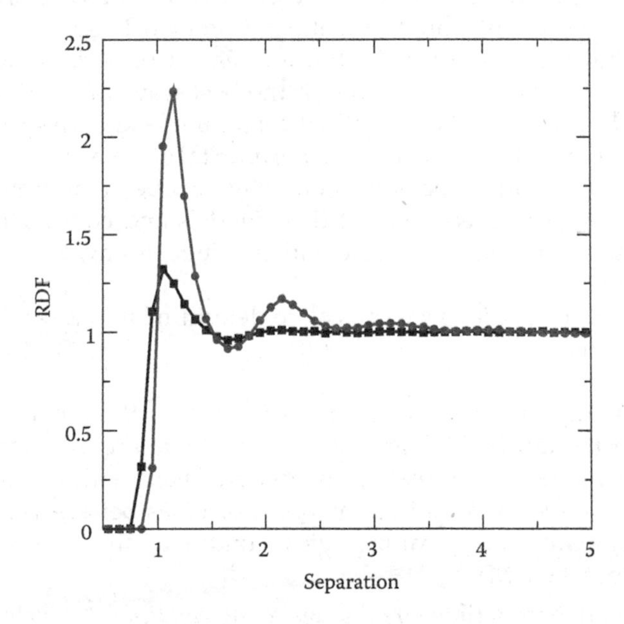

FIGURE 18.2
Plots of the radial distribution function (RDF), which measures the correlation between particles as a function of separation, normalized relative to an ideal gas. A gas phase system is observed for $\epsilon = 0.1$ (black curve), as indicated by the lack of long-range correlations at intermediate and long-range separations (i.e., lack of peaks in the curve for separations greater than 2σ). A liquid phase is seen for $\epsilon = 1.0$ (gray curve), as indicated by the presence of correlations at intermediate separations (i.e., the peak at around 2σ). Simulations were performed at $T = 1.0$, with $\sigma = 1.0$, and reported in reduced units using LAMMPS.

be very flexible when visualizing a simulation trajectory, we could immediately identify an issue in the model, without the need for further validation tests.

In a related context, pushing the limits of the model allows one to identify other behaviors that may be problematic. For example, parameters used to describe ϵ for the Lennard-Jones potential are typically reported in the literature in units kcal/mol or kJ/mol; the LAMMPS molecular dynamics simulation engine expects parameters to be defined in kcal/mol (for "real" units). As a result of the similar order of magnitude (roughly a factor of 4 difference between kcal/mol and kJ/mol), LAMMPS simulations with incorrect kJ/mol units will often run to completion with no obvious errors (i.e., the equations of motions can be stably integrated), although significantly different phenomenological behavior is likely to be observed. That is, we might assume bad parameters would lead to an obvious "crash," or even failure of basic sanity tests, but this must be evaluated on a case-by-case basis. As such, it is often useful to induce errors to understand how sensitive the model is and understand how it will fail. For example, if an order of magnitude change in a parameter has negligible effects on the output of the model, but a small variation in another parameter dramatically changes the behavior, it becomes clear not only which parameter should be the focus when it comes to model refinement but also which parameters, if incorrect, may not produce obvious errors. Inducing errors provides a clear cause and effect, which can make it easier to identify the root cause of problems that may occur later on during usage of a model.

18.4.3 Comparisons with Existing Results

If we now consider the inputs to be accurately defined, we must determine if the model itself produces the correct behavior. As mentioned previously, sanity checks can help us eliminate models that are clearly wrong, but it does not provide us with knowledge of whether the model is actually correct. One of the best ways of validating a model is to compare its results to a "known" system. What constitutes a known system has evolved in recent decades from purely theoretical and experimental systems to now include properly curated simulations. For many problems, there may now be an abundance of choices for sources against which researchers can test their models and codes. However, given this proliferation of results, how can we decide which results, if any, are actually suitable for validation purposes?

There are a number of questions we can ask to determine how useful an existing result will be for comparisons.

- First, and most importantly, *how similar are the systems that are being compared?* For instance, even though both diamond and graphene are allomorphs of carbon, results for one model are unlikely to be in agreement with results for the other. Similarly, it is well known that various biomolecules behave very differently in pure water compared to typical biological conditions (in a salt solution of ionic strength of about 0.150 M).
- If the systems are compatible, *are the "operating conditions" compatible?* It may be possible to use a model developed for a liquid at 300 K for a gas phase at 1000 K to compare with someone else's results at 1000 K, but such comparisons make sense only if the computational model is known to be valid for a gas at 1000 K (i.e., models may or may not be fully transferable to different thermodynamic statepoints).

- *What level of accuracy can be expected in the comparison?* If the systems are quite disparate—for instance, due to computational cost, atomistic simulation models typically consider shorter polymers than what can be realized experimentally—then we may only be able to achieve qualitative agreement of trends between the two sets of results.
- *What properties does the model need to reproduce?* If we are only interested in the behavior of a solid structure, it would be less important to validate that the model accurately reproduces the gaseous state. Furthermore, if we were to validate against only the gaseous state, and not the solid state, such validation would essentially be meaningless.

When considering both experimental and simulation comparisons, we should also ask if the comparisons are themselves methodologically sound. This is particularly a concern for older results from both domains. Experimental results may have been overturned by new evidence or new measurement techniques or because prior protocols were determined to be flawed. It is also possible that the experimental protocol can directly affect the conclusions of a paper in subtle ways: for instance, if raw materials are supplied by different sources, resulting differences could be ascribed to the differences in material preparation and purity rather than an intrinsic physical or material property being investigated. Similarly, there may be inherent issues with older simulation results, as a result of small system sizes, insufficient convergence, or other "shortcuts" applied to overcome memory or time limitations that were more constricting in the past than they are today. More significantly, such results may have been produced with "in-house" codes that are no longer available, compilable, or executable and may not have been fully verified, making it difficult to ascertain the root causes of any discrepancies that may arise.

As an example of validating a model through comparison, consider validating the ReaxFF [14] forcefield as a model for the elongation of gold nanowires. While the ReaxFF forcefield was validated against bulk crystals and minimal energy clusters in prior studies [15], the behavior of nanowires was not explored. Thus, before using the forcefield to study nanowires, we should conduct *relevant* validation on the specific system of interest. In Ref. [16], computationally expensive first principles density functional theory (DFT) calculations were performed for small gold nanowires (diameter, D = 1.1, 1.5, and 1.9 nm) and several configurations representing nanowires undergoing mechanical deformation. These configurations were used as input to ReaxFF and the energy evaluated and compared to DFT.

Figure 18.3 plots the DFT energy as a function of ReaxFF energy for identical structures, where in this plotting scheme, an ideal match occurs when a point lies on the line $y = x$. ReaxFF predicts near-perfect agreement with DFT, both in terms of quantitative values and also the trend of the data, where a least squares regression of the data yields a slope of 0.99 (an ideal match would be a slope of 1.0).

As a further comparison, the same procedure was carried out using the tight-binding second-moment approximation (TB-SMA) forcefield [17], also shown in Figure 18.3. For TB-SMA, more significant deviations from the DFT energy are seen and the slope of the data is 1.63, thus demonstrating failure to provide either qualitative or quantitative agreement. Thus, we would conclude that, given these data, ReaxFF is validated for these types of nanowires and TB-SMA fails validation. It is important to note that, if we instead validated the models by relying on available data for bulk crystals, rather than performing calculations on the system of interest, that is, nanowires, both ReaxFF and TB-SMA would pass validation, as both accurately reproduce bulk behavior [15,17], clearly highlighting the importance of choosing the appropriate data for validation.

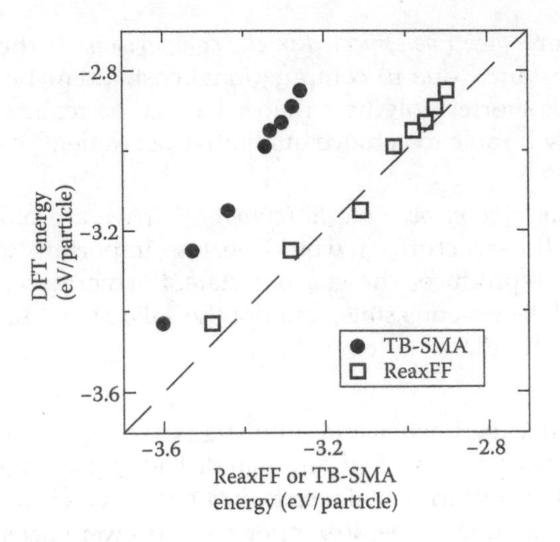

FIGURE 18.3
DFT calculations of the energy of small nanowires (diameter, D = 1.1, 1.5, and 1.9 nm) and several elongated structures, plotted against energy calculated for the same structures using the ReaxFF and TB-SMA forcefields. Ideal matching corresponds to data points that lie along the line $y = x$, where ReaxFF shows near perfect agreement with DFT. (Adapted from Iacovella, C.R., French, W.R., Cook, B.G., Kent, P.R.C., and Cummings, P.T., *ACS Nano*, 5, 10065–10073, 2011.)

18.5 Summary

A few main items to remember when considering validation processes for computational models and codes include:

- Successful execution of a code does *not* guarantee that the results of the code are correct for a given problem.
- Proper validation requires a "well sourced" set of data, that is, both accurate and appropriate, which can be used to check the results obtained from a model.
- Errors can be introduced by either choosing the wrong model to solve a problem or providing incorrect values for parameters that are passed as inputs to the model.
- Validation should be completed *before* production runs start, if at all possible.
- Creating adequate documentation and using/developing automated tools to perform sanity checks for input files can reduce errors and minimize the time and effort required for debugging and troubleshooting.

References

1. Arthur G. Stephenson, Daniel R. Mulville, Frank H. Bauer, Greg A. Duke-man, Peter Norvig, Lia S. LaPiana, Peter J. Rutledge, David Folta, and Robert Sackheim. *Mars Climate Orbiter Mishap Investigation Board: Phase I Report*. NASA: Technical Report. November 1999.

2. http://mars.jpl.nasa.gov/msp98/orbiter/.
3. Yongsheng Leng, Predrag S. Krstić, Jack C. Wells, Peter T. Cummings, and David J. Dean. Interaction between benzenedithiolate and gold: Classical force field for chemical bonding. *Journal of Chemical Physics*, 122(24):24471, 2005.
4. Peter Pivonka and Svetlana V. Komarova. Mathematical modeling in bone biology: From intracellular signaling to tissue mechanics. *Bone*, 47(2):181–189, 2010.
5. Lan Wang, Cheng Li, Guinevere Kauffmann, and Gabriella De Lucia. Modelling galaxy clustering in a high-resolution simulation of structure formation. *Monthly Notices of the Royal Astronomical Society*, 371(2):537–547, 2006.
6. Ozan K. Tonguz, Wantanee Viriyasitavat, and Bai Fan. Modeling urban traffic: A cellular automata approach. *IEEE Communications Magazine*, 47(5):142–150, 2009.
7. https://cdmhub.org.
8. Daan Frenkel and Berend Smit. *Understanding Molecular Simulation: From Algorithms to Applications*. San Diego: Academic Press, 2002.
9. M.P. Allen and D.J. Tildesley. *Computer Simulation of Liquids*. Oxford, England, New York: Clarendon Press, 1989.
10. Steve Plimpton. Fast parallel algorithms for short-range molecular dynamics. *Journal of Computational Physics*, 117(1):1–19, 1995. http://lammps.sandia.gov.
11. Joshua A. Anderson, Christian D. Lorenz, and A. Travesset. General purpose molecular dynamics simulations fully implemented on graphics processing units. *Journal of Computational Physics*, 227(10):5342–5359, 2008. http://codeblue.umich.edu/hoomd-blue.
12. William Humphrey, Andrew Dalke, and Klaus Schulten. VMD: Visual molecular dynamics. *Journal of Molecular Graphics*, 14(1):33–38, 1996.
13. Sander Pronk, Szilárd Páll, Roland Schulz, Per Larsson, Pär Bjelkmar, Rossen Apostolov, Michael R. Shirts et al. GROMACS 4.5: A high-throughput and highly parallel open source molecular simulation toolkit. *Bioinformatics (Oxford, England)*, 29(7):845–854, 2013.
14. Adri C.T. van Duin, Siddharth Dasgupta, Francois Lorant, and William A. Goddard. ReaxFF: A reactive force field for hydrocarbons. *Journal of Physical Chemistry A*, 105(41):9396–9409, 2001.
15. John A. Keith, Donato Fantauzzi, Timo Jacob, and Adri C.T. Van Duin. Reactive forcefield for simulating gold surfaces and nanoparticles. *Physical Review B, Condensed Matter and Materials Physics*, 81(23), 2010.
16. Christopher R. Iacovella, William R. French, Brandon G. Cook, Paul R.C. Kent, and Peter T. Cummings. Role of polytetrahedral structures in the elongation and rupture of gold nanowires. *ACS Nano*, 5(12):10065–10073, 2011.
17. Fabrizio Cleri and Vittorio Rosato. Tight-binding potentials for transition metals and alloys. *Physical Review B, Condensed Matter and Materials Physics*, 48(1):22–33, 1993.

19

Software Licensing and Distribution

Paul Saxe

The previous chapters in this book have covered the many aspects that you need to think about as you develop code to solve the problem at hand, how to make it perform well on current machines, and so on. In this chapter, we will make a complete change of direction and assume that you have written a useful and valuable code. You have tested it and found that it does what you want, and you are proud of what you have accomplished. Now what? If you want other people to use your code, you can post it on your web site, e-mail it to friends and colleagues, post a note on a mail list, or talk about it at a conference. And people will start asking you if they can have a copy of your code. That is wonderful! Your work will help others, and they may add to it and make it better.

However, before you let your new creation out into the world, it is worth stopping and thinking about what you want to accomplish and what will happen as you distribute your code. This is where licensing comes in. How you license your code will determine what others can—and cannot—do with it. You may well be thinking that this sounds like a lot of legal mumbo-jumbo that is not worth the effort and besides is not needed. After all, you are just going to give your code to some people who you know and trust and they will be careful to recognize your work. Right? The reality is that you don't actually know what is going to happen in the future. What if the people that you give the code to give it to other people who don't know you? What if the code grows into a major code over time, a code that is useful to many people? It's a good idea to think about what you want to achieve now, before you start distributing your code. When you license your code, you set the conditions under which others can use the code. You can find or create a license that says almost anything you want—it can apply many restrictions or almost none at all—but it sets a framework so that there are no accidental misunderstandings or people doing things with your code that you don't want. Licensing your code is much like buying insurance for your car or house. It may not be something you enjoy doing, and you hope that you will not ever need it. But it is something you should do, something that can prevent major problems down the road. And unlike home insurance, it doesn't cost you anything except a small amount of time.

Before we go further, there is a necessary disclaimer: the author of this chapter is not a lawyer and this is not legal advice. It is not complete and may not be fully correct, so please consult your own lawyer for questions about licensing your code. The intent of this chapter is to give you an introduction to licenses and prompt you to think about some of the issues, but the rest is up to you.

Also, to make sure it is clear, you are licensing the code itself, not the results that it produces. When you license the code, which means either the source code or the binary executables generated from the source, the license is based on copyright laws. Such laws do not apply to the output from the program. You do not generate the output, the user does, so it is theirs, not yours.

First, we will cover the general types of licenses available. You can of course develop your own license if you wish, but a good place to start is understanding the licenses that are available. There are two main types of licenses—proprietary and open source—which differ mainly in whether and how the end user is allowed to further distribute the code. A proprietary license, which is the type used by most commercial codes, typically does not let the end user give the code to anyone else. An open source license allows the end user to further distribute the software; however, there are again two general types of open source licenses. Permissive licenses do not place many restrictions on redistributing the code, allowing an end user to distribute the software under a different license. This allows, for example, software under a permissive license to be incorporated in, for example, commercial software. Copyleft licenses, sometimes also called "viral licenses," are more restrictive because they try to ensure that anyone who receives the software has the same rights as anyone. In order to do this, copyleft licenses require that the software be redistributed under the same license and that any modifications or additions also be distributed under the same license. This implies that software licensed under a copyleft license cannot be incorporated into commercial software, for example, because the entire work would need to be licensed under the copyleft license and hence distributed freely. There are some partial copyleft licenses that weaken the restriction for additions to the code, allowing them to be distributed under a different license. The original code and modifications to it must be redistributed under the original license. In a sense, the partial copyleft licenses are between the permissive licenses and the full copyleft licenses. If you create a larger work by adding separate parts to the software, then you may license those as you wish, much as you could with a permissive license. However, modifications to the original code must be released just as in the copyleft licenses.

Finally, it is worth mentioning that if you distribute software but do not license it, you are effectively allowing anyone to do whatever they wish with it. Technically, you still hold a copyright on the software, but if you do not include a notice or license, users may not even realize that you wrote the software. Effectively, though perhaps not legally, it enters the public domain. Alternatively you can use the CC0* license from Creative Commons to legally relinquish as much of your copyright and related rights as possible.

A key concept implied in the discussion above is the compatibility of licenses. The permissive licenses are typically compatible with most other licenses; that is, you can mix code licensed under a permissive license with code licensed under another type of license because the permissive license allows you to relicense under the different license. The copyleft and proprietary licenses typically do not allow mixing with code under different licenses since that would defeat their purpose. The partial copyleft licenses are, as before, somewhere in between. The compatibility between licenses can, however, depend on some quite small details in the licenses, so it is difficult in general to say whether licenses are compatible. Fortunately, however, some of the major licenses list which other licenses they are compatible with (Table 19.1).

Why would you want to use one of these licenses rather than another? Or one of the many other licenses available? That depends on what the software that you are writing does, how long you think it might last, and what your goals are.

If your code is short, say no more than a few hundred lines of code, unless it does something quite exceptional, it is probably not worth much effort thinking about the licensing

* Creative Commons CC0 license, https://creativecommons.org/about/cc0, accessed October 14, 2015.

TABLE 19.1

Common, Widely Used Open Source Licenses

License	Type	GPL Compatible
GNU Public Library v3 (GPL3)[a]	Copyleft	Yes
Lesser GPL v3 (LGPL3)[b]	Copyleft	Yes
Mozilla Public License v2 (MPL2)[c]	Partial copyleft	Yes
Eclipse Public License v2 (EPL2)[d]	Partial copyleft	No
Common Development and Distribution License (CDDL 1.0)[e]	Partial copyleft	No
Apache License v2[f]	Permissive	Yes
BSD 3 clause[g]	Permissive	Yes
BSD 2 clause[h]	Permissive	Yes
MIT License[i]	Permissive	Yes
Creative Commons CC0 license[j]	Public domain	Yes

[a] The GNU General Public License, version 3, http://www.gnu.org/licenses/gpl-3.0.en.html, accessed October 14, 2015.
[b] The GNU Lesser Public License, version 3, http://www.gnu.org/licenses/lgpl.html, accessed October 14, 2015.
[c] The Mozilla Public License, version 2, https://www.mozilla.org/en-US/MPL/2.0/, accessed October 14, 2015.
[d] The Eclipse Public License, https://www.eclipse.org/legal/epl-v10.html, accessed October 14, 2015.
[e] The Common Development and Distribution License, http://opensource.org/licenses/CDDL-1.0, accessed October 14, 2015.
[f] The Apache License, version 2, http://www.apache.org/licenses/LICENSE-2.0, accessed October 14, 2015.
[g] The BSD 3-Clause License, https://opensource.org/licenses/BSD-3-Clause, accessed October 14, 2015.
[h] The BSD 2-Clause License, https://opensource.org/licenses/BSD-2-Clause, accessed October 14, 2015.
[i] The MIT License, https://opensource.org/licenses/MIT, accessed October 14, 2015.
[j] Creative Commons CC0 license, https://creativecommons.org/about/cc0, accessed October 14, 2015.

or in adding the license. Think about placing it under one of the permissive licenses, perhaps the Apache License, version 2. This is the recommendation of the Free Software Foundation,* who strongly recommend the copyleft GPL for most software. You could also use any of the other permissive licenses; however, the Apache License is more recent and includes language covering patented software, which the other common permissive licenses do not since they were written before software patents became possible.

If the software that you have developed is more substantial than a few hundred lines, you need to think carefully about the license and your goals and how long you want to be involved with the software. The motivation behind copyleft is that software should be free and open: "the users have the freedom to run, copy, distribute, change and improve the software."† This is the reason that the copyleft licenses insist that the license apply to any modifications or additions to the software that are redistributed. If you use copyleft software as part of your software, then you have no choice but to license your software—it is a "derivative" work—using a compatible copyleft license. In practice, this means that if you incorporate GPL-licensed code into yours, either directly as source code or by linking to the GPL code, you need to use the GPL as the license. This also means that if you use a strong copyleft license like the GPL, then anyone who modifies and adds to your code must distribute their version under the same license, typically the GPL.

* Frequently Asked Questions about the GNU License. What if the work is not very long? http://www.gnu.org/licenses/gpl-faq.html#WhatIfWorkIsShort, accessed October 14, 2015.
† What Is Free Software? http://www.gnu.org/philosophy/free-sw.html, accessed October 14, 2015.

This approach maximizes the freedom of all the individuals in the community to work with the code, to extend it, and so on. At the same time, it prevents commercialization of the code itself, although it does not prevent a commercial company charging for installation, support, training, and other activities around the software. This is indeed the business model of successful companies like RedHat and SUSE, building around the open source Linux operating system. Clearly, such an approach works well when there is a large, active community of users who can contribute back to the project. It is less obvious how well it works with a small community of users, where it may be hard to find enough users to support and develop the code. If you are thinking about releasing your code under a copyleft license like the GPL, and you expect that the code will last for a long time, you should think about whether you intend to maintain and develop the code for that long time or if there will be a sufficient community to take it over. If the community is small, and there are only a couple people actively supporting the code, it is possible that they will change jobs or no longer be able to work on the code, in which case it might die for lack of support. This happens all the time—there is a natural selection of software going on—which is not a problem unless you think your software should continue to live. If the community is small, you may be the main supporter of the code for a long time.

Permissive open source licenses approach the freedom of users in a different way. They provide almost complete freedom to the initial end users but do not guarantee that subsequent users will have the same rights. Permissive licenses do this by allowing you to redistribute the software under a different license, which can be anything from a copyleft license such as the GPL to a proprietary, commercial license. A user can modify and add to the code but is not forced to give it back to the community. This is why the licenses are called "permissive"—a user can do almost anything with the code, including incorporating it into a commercial product. Of course, you might not like someone else taking your code and profiting from it, or improving it but not giving the changes back to you. However, if the community is small, and your software has commercial value, using a permissive license does open another avenue if you want the code to continue living but you cannot or do not want to continue working on it.

Partial copyleft licenses, like the Mozilla Public License (MPL-2) and the LGPL, split the difference. If you use such a license, your code can be incorporated into a larger work that might be under a different license, but your code and modifications to your code remain under a copyleft license where the changes must be made available to the entire community. Typically, these licenses work at the level of source files, so any changes in the source files that you distribute remain public, but added files of source may not. The LGPL operates at the level of libraries, allowing an LGPL'ed library to be linked with code licensed differently. As with the permissive licenses, this increases the community of supporters for your software to include commercial software companies, although they would be obliged to publish modifications to your code, which is not the case with permissive licenses.

The final type of license, the proprietary license, is self-explanatory. You maintain as much control over your software as you wish and can forbid end users from copying the software. This is the typical approach of commercial companies, who charge for the software and use that revenue to support and maintain the code, as well as for profits, and others.

At several points in the discussion above about the open source licenses, there have been references to whether the community is "large" or "small." It is not easy to define what "large" or "small" means in this context. A large community simply means one large enough, with enough active users, to support the continued development and maintenance of a code. This will depend greatly on the complexity of the code, the types of users, and their motivation. If you are thinking about using an open source license, this is the key question that you have to answer when deciding what type of open source license to use. If you choose a copyleft license and the community is too small, your code will live as long as you work on it. If you choose a permissive license and the community is large, you may miss opportunities to benefit from the work of others in the community.

You are not subject to your own license: it applies to end users. Therefore, you can license your software in multiple ways, and you can change the license for your software. However, you cannot take back a license, so if you have licensed your software under an open source license and decide to change the license to a commercial license, then the last version that you released under the open source license will remain available, at least through other end users. If you have incorporated others' code into yours or have incorporated improvements that others made back into it your code, you cannot change the license without their permission. If many people have made changes to the code, it may be difficult to contact them, let alone get everyone's permission to change the license. Thus, for large, complicated codes that have had many contributors, it becomes difficult, if not practically impossible, to change the license. This is one of the reasons to be careful to choose the right license early.

Hopefully, this chapter has given you some ideas and enough understanding to decide on the type of license to use for your software. Open source does not mean just the GPL! Many people seem to think that GPL is open source and open source is GPL, but there are many other licenses available, one of which might suit you better. While GPL and similar licenses have been very useful and effective where there is a large community, much scientific and technical software has a smaller audience. While universities and national labs are creating many open source codes, this is often based on funding from grants. This is a very different situation to the large open source projects such as Linux. If a code is dependent on grants for funding, what will happen when the funding shifts? This may be an issue with the copyleft licenses, particularly for large codes with many contributors. If the funding dries up but the community is not large enough to support the code without grants, there may be no way to maintain the code, and with many contributors, it may not be possible to change the license. Were the software released under a permissive open source license instead, and is of enough utility, it might be possible to fund it commercially, either by selling it directly, by starting a software company, or by working with an established software company, thus providing possible options for growing the software and making it part of comprehensive software solutions.

As a final note, the institution that you work for may have rules about licensing software, as may funding agencies. For example, software developed by employees of the U.S. government is automatically in the public domain and cannot be licensed. Also, the European Commission has released documents about software development and licensing that you may find useful.* You need to check with your institution and also understand any requirements connected with grants and other funding sources to make sure that you abide by their policies with respect to software and licensing. An excellent source for further information on open source and various licenses is the Open Source Initiative.†

* The European Materials Modelling Council (EMMC) Guidance on Quality Assurance for Academic Materials modelling software engineering, http://emmc.info/wp-content/uploads/2015/06/Quality-of-Materials-Modelling-Software_V3_201502264.pdf, accessed October 14, 2015.

† Open Source Initiative, http://opensource.org, accessed October 14, 2015.

Index

H

P

Q

R

S

W

X

Z